Professional Reference for Teachers

HOLT, RINEHART AND WINSTON

A Harcourt Classroom Education Company

Austin • New York • Orlando • Atlanta • San Francisco • Boston • Dallas • Toronto • London

To the Teacher

In this *Professional Reference for Teachers,* you will find words of wisdom from your colleagues in the field of education.

Section I describes in detail the relevance, practical application, and value of such classroom strategies as cooperative learning, multicultural instruction, and concept mapping.

Section II is a collection of articles from teachers, professors, and policy makers. Turn to this section for practical expertise on topics such as effective classroom management, dealing with aggression in the classroom, ensuring the success of girls in science, dispelling the "scientist as nerd" myth, involving parents in their child's study of science, and implementing the National Science Standards.

Section III contains a bank of additional resources. Here you will find a listing of other works by authors included in Section II. In addition, lists of related books, periodicals, and audiovisuals as well as the names and contact information for organizations and associations are available for your reference. A complete index is also available to help you access specific topics with ease.

We'd Appreciate Your Opinion!

At the very back of this book you will find an Opinion Form. What is your opinion of the *Holt Science and Technology* program? We at Holt, Rinehart and Winston would like to know! An electronic version of the form is available on the *One-Stop Planner CD-ROM*. Or you can send your responses directly to us via E-mail by accessing the form on our Web site at http://go.hrw.com/surveys/hst

Art/Photo Credits
All work, unless otherwise noted, contributed by Holt, Rinehart and Winston. Abbreviated as follows: (t) top; (b) bottom; (c) center; (bkgd) background.

Cover: (tl), Jose L. Pelaez/Stock Market; (tr), Image Copyright ©2001 Photodisc, Inc.; (bl), EyeWire, Inc.; (bc), EyeWire, Inc.; (br), Image Copyright ©2001 PhotoDisc, Inc.

Additional credits: See page 155.

Printed in the United States of America

ISBN 0-03-054422-X
1 2 3 4 5 6 032 04 03 02 01 00

Contents

Teaching Science

Professional Articles

Continuing the Discussion

Science, Technology, and Society (STS)

Science and technology are flip sides of the same coin; each supports the other. Exploring the relationship between science and technology, one will discover that both affect our society as a whole. Even people who never again set foot in a laboratory after leaving school can benefit from an understanding of science and technology and how both relate to each other and to society at large.

In our society, complex as it is, this fact is now more true than ever before. Science and technology will play roles in every aspect of life. In the future, "high-tech" will be more than a catch phrase—it will permeate every aspect of life. To prepare students for the challenges of the future, they must become science literate. They must be given the tools that will enable them to become responsible and productive individuals in a highly technological world.

STS is an approach to teaching science in which the impacts of scientific, technological, and social matters are explored. STS teaches science from the context of the human experience and in so doing leads students to think of science as a social endeavor. STS emphasizes personal involvement in science. Students become active participants in the scientific experience.

There are three parts to STS:

sts **Science concept and skill development, and knowledge of the nature of science** This component of STS introduces science as a system for learning about the natural world and gives students the foundation they need to actually practice science in and out of the classroom. The ultimate goal of STS is to turn students into scientists, at least for the duration of their science education. To accomplish this, students first learn the methods of science and the skills that scientists draw on. Whenever possible, the major ideas of science are introduced from the standpoint of those who developed them. In this way, students come to see the reasoning that went into the development of these ideas.

sts **Knowledge of the relationship of science and technology, and engagement in science-based problem solving** Students who understand the real-world applications of science are better able to appreciate and enjoy it. This component of STS reinforces the practical value of science. Students see that science is a system for solving practical problems. Students themselves become practical scientists—first identifying problems and then developing solutions to them. Students learn to analyze, to plan, to organize, to design, and to refine models and designs.

sts **Engagement in science-related social issues and attention to science as a social institution** This component of STS deals with the ways in which science serves human needs. The benefits may be tangible or intangible, but either way they are real. To emphasize the social responsibility of science, scientists are shown to be concerned about the impact of their work on society as a whole. Advances in science and technology sometimes lead to thorny ethical issues. Such issues are often used as a focus for discussion and investigation. From these investigations, students draw conclusions and form reasoned opinions.

STS emphasizes personal involvement in science. Students become active participants in the scientific experience.

By strengthening your students' understanding of the connections between science, technology, and society, you will enhance their appreciation and understanding of science and their ability to deal sensibly with the complex issues that will become commonplace in the decades ahead.

Themes in Science

Too often, students view science as a system of separate, unrelated abstractions or as a compilation of facts and difficult-sounding terms. But science is simply the study of nature, and there are certain underlying principles, or themes, that unite the study of all areas of science. The themes provided here are not meant to replace the traditional teaching of scientific disciplines but rather to help teachers create a framework for the unification of these disciplines.

The following themes can be used with Holt, Rinehart and Winston's middle school science programs. These themes are intended to meld facts and ideas and to provide a context for discussing the textual matter in a meaningful way. You can employ these themes as an organizational tool or to reinforce understanding of the subject matter.

Energy

Energy puts matter into motion and causes it to change. Energy is what makes the universe and everything in it dynamic. The study of dynamic systems in any field of science requires an understanding of energy: its origins, how it flows through systems, how it is converted from one form to another, and how it is conserved. Energy provides the basis for all interactions, whether biological, chemical, or physical. Thematically, energy connects all scientific disciplines.

Systems

A system is any collection of objects that influence one another. The parts of a system can be almost anything—planets, organisms, or machines, for example. A system may be very small, such as a cell nucleus; very large, such as a galaxy; or very complex, such as the human body. All of the scientific disciplines involve the study of some kind of system or systems. Understanding a system involves knowing what its important parts are and how those parts work together.

Structures

Structure provides a basis for studying all matter, from the most basic forms to the most complex. Structure is closely related to function, so scientists study the structures of things to learn how they work. For example, the structure of an eye reveals much about the process of vision. And the key to a diamond's strength lies in the tight lattice structure of carbon atoms within the diamond.

Changes over Time

A change over time is not a single alteration, but a progression of alterations that occurs over the continuum of time. Biological evolution is an example of change over time. Evolution has gradually changed the characteristics of organisms ranging from the starfish to the giraffe. Changes over time occur in physical and Earth science as well. For example, the chemical change of rust forming on an iron nail can be traced through time, as can the movements of the Earth's plates. An understanding of how changes occur over time allows students to appreciate not only the present state of the world around them but also what it may have looked like in the past and what it may look like in the future.

There are certain underlying principles, or themes, that unite the study of all areas of science.

Cycles

A cycle is a pattern of events that recurs regularly over time or a circular flow of materials in a system. Cycles occur throughout nature and appear in all of the scientific disciplines. The water cycle, the rock cycle, and the life cycle of a particular organism are examples of cycles at work. A defining feature of a cycle is that it has no beginning or end; therefore, the study of any part of a cycle is incomplete without consideration of the other elements in the cycle.

In your class discussions, use the themes to provide a framework of understanding for your students.

Using Themes

In your class discussions, use themes to provide a framework of understanding for your students. For example, whether you are studying photosynthesis or the way in which the forces of nature have shaped the physical appearance of the Earth, the theme of **Energy** can be discussed. Similarly, one subject can be addressed from the viewpoint of many different themes. For example, in discussing an organism such as a zebra, **Energy** can be applied in a discussion of how the zebra takes in food from the environment; **Changes over Time** can be discussed in relation to how the zebra's structures are adaptations to its environment; and **Cycles** can be introduced in a discussion of how the zebra's migration habits are based on the seasons.

Journals and Portfolios

Keeping a Journal

One highly successful tool for improving students' performance in science is the journal. This program's name for the journal is the ScienceLog. The ScienceLog has many functions. First and foremost, it is an ongoing record of students' learning. Students begin the study of a new topic by recording prior knowledge of that topic. Any misconceptions that students may have are thus exposed. As the lesson progresses, students record any and all new findings. In many cases, students find that what they learned in the activities contradicts their preconceptions.

Much of the work that students do should be recorded in their ScienceLog. The ScienceLog is a constant reminder to students that learning is occurring. Students can look back and compare their early work with later work to see and take pride in the progress they have made. To supplement their other work, you may want to ask students to briefly summarize what they have learned each week. This makes a very handy capsule history of their work.

What makes a good ScienceLog? Insofar as is possible, the ScienceLog should be neat and easy to follow. Students may organize their ScienceLog in any of a number of ways—chronologically, by chapter or unit, or by lesson, to name a few. Some kind of heading should distinguish each major entry. A spiral-bound notebook or hard-bound lab-type notebook makes a good ScienceLog.

Portfolios can be elaborate collections of work that are limited only by the teacher's and students' imagination.

Portfolios: What, Why, and How

Definitions and descriptions vary as to what constitutes a portfolio. Very simply, a portfolio is a collection of work that is done by the student during the course of the year. Usually, the students themselves have a say as to what goes into the portfolio, but selections should represent the objectives outlined by the curriculum. The purpose of the portfolio is to represent the students' mastery of skills and knowledge within the subject area. This may sound simple enough; in fact, it can be very simple. On the other hand, portfolios can be elaborate collections of work that are limited only by the teacher's and students' imagination.

Teacher to Teacher

Find out what your colleagues have to say about this hot topic in their article "Assessment that Emphasizes Learning" on page 106.

Your initial decision to use portfolios in your assessment leads to a host of other decisions that must be made. Before you launch into making these decisions, it is a good idea to seek input from other teachers as well as from your school administration. You may even want to discuss your ideas with parents and other members of your community. This will make it easier to get feedback later as to the impact portfolios are having on student learning and attitudes.

You will also need to plan a system for organizing and managing the portfolios. Establish guidelines for what types of information will be admitted into the portfolios, how and by whom the selections will be made, and when materials can be added or revised. These decisions will be based on your individual preferences and on the level and attitudes of your students. For help developing your plans, see the guidelines on the next page.

Portfolio Guidelines

- Allow students to select for themselves the sample materials that best represent their level of understanding and mastery. Although you may require that certain projects or materials be placed in the portfolios, it is advised that the students play an active role in selecting their best work. This gives students ownership and encourages them to take increasing responsibility for the quantity and quality of their work.

- Allow students to revise the selections in their portfolios at any time. The portfolios should evolve as your students' skills improve.

- Identify well in advance when you will be assessing the portfolios. Students should be given plenty of warning before the portfolios are collected for assessment.

- Determine in advance the criteria you will use for grading the portfolios. Share this criteria with the students up front. You may want to make a criteria checklist that each student can place inside his or her portfolio. This checklist would provide both you and your students with a ready reference to the grading criteria.

- You may want to keep the portfolios in the classroom or in some other area, allowing frequent but controlled access to them. This will decrease the chances of portfolios being lost. This may require a significant amount of room, depending on the nature of the portfolios.

- Make available examples of high-quality portfolios so that students can see examples of excellent work. This, of course, may be possible only after you have used portfolio assessment for at least one course.

Portfolio Links in this Program

The ScienceLog provides an excellent opportunity for portfolio assessment. By directing your students to include in their ScienceLog representative samples of their work, as well as comments and reflections about their samples, you can add depth to your evaluation method. In addition, to help you use Portfolios effectively, specific strategies are provided in Portfolio boxes in the margins of this program's Annotated Teacher's Edition.

Concept Mapping

> **Concept maps are a highly effective tool for helping students make logical connections.**

Too often, students are able to master the individual elements of a topic without truly grasping the "big picture." If students fail to understand how the elements fit together or relate to one another, they cannot truly comprehend the topic. Concept mapping is a very effective method of helping students see how individual ideas or elements connect to form a larger whole. Concept maps are a highly effective tool for helping students make those logical connections.

The most effective concept maps are those that students construct on their own. Used in this way, concept maps are both a self-teaching system and a diagnostic tool. To construct a proper concept map, the student must first examine closely his or her mental model of the topic at hand. Any flaws or shortcomings in that model will be reflected in the concept map.

Concept maps are flexible. They can be simple or highly detailed, linear or branched, hierarchical or cross-linked, or they can contain all of these major elements. Students can construct their own maps from scratch or can finish incomplete maps. Concept maps can take almost any form as long as they are logically arranged.

Making Concept Maps

The steps involved in making a concept map are outlined below.

1. Make a list of the concepts to be mapped. Concepts are signified by a noun or short phrase equivalent to a noun.
2. Choose the most general, or the main, idea. Write it down and circle it.
3. Select the concept most directly related to the main idea. Place it underneath the main idea and circle it. If two or more concepts bear the same relationship to the main idea, they should be placed at the same level.
4. Draw a line between the related concepts, leaving a space for a short action phrase that shows how the concepts are related. These are linkages.
5. Continue in this way until every concept in the list is accounted for.

The simple concept map below shows the relationship among the following terms: plants, photosynthesis, carbon dioxide, water, and sun's energy. More detailed maps are shown on the next page.

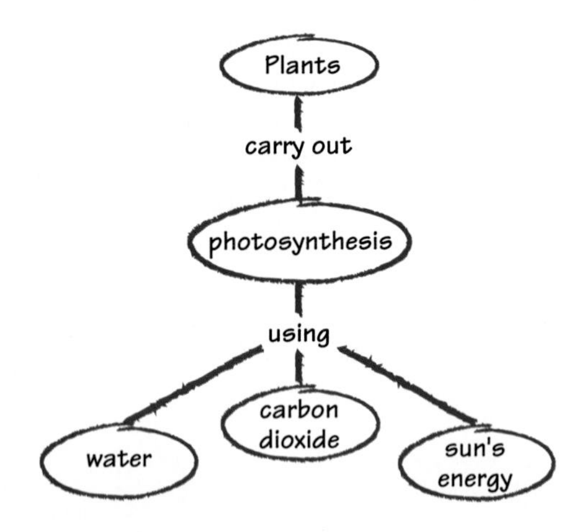

For any given topic, there is no single "correct" concept map. Not all maps are equally valid, however. Good concept maps have most or all of the following characteristics:

- start with a single, general concept—a big idea—and work down to more specific ideas
- represent each concept with a noun or short phrase, each of which appears only once
- link concepts with linkage words or short phrases
- show cross-linkages where appropriate
- consist of more than a single path
- include examples where appropriate

Using Concept Maps

Concept maps can be applied in many ways, such as the following:

- to gauge prior knowledge of a topic
- as an end-of-lesson, chapter, or unit evaluation
- as a pretest review
- to help summarize special presentations, such as films, videos, or guest speakers
- as an aid to note taking
- for reteaching

You may also want to use partially completed concept maps as pop quizzes or as devices for summarizing particularly difficult class sessions.

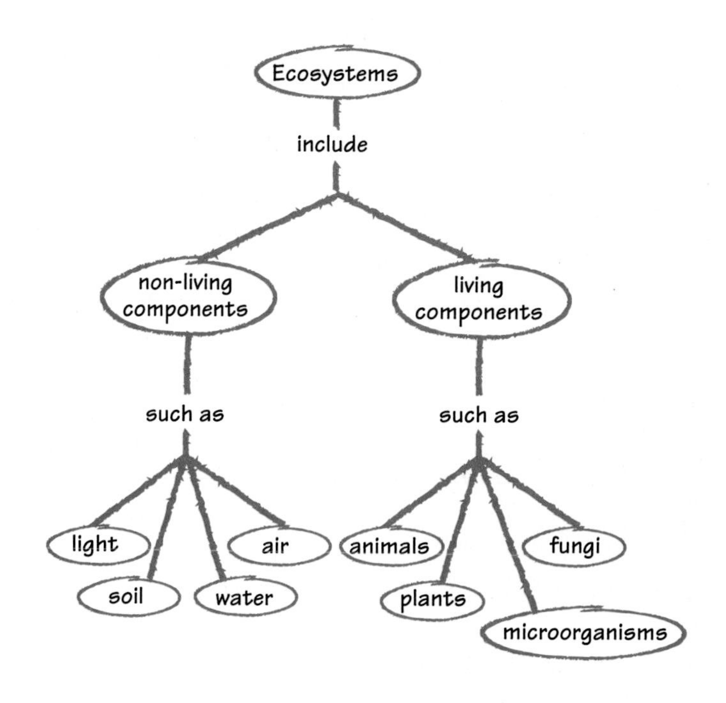

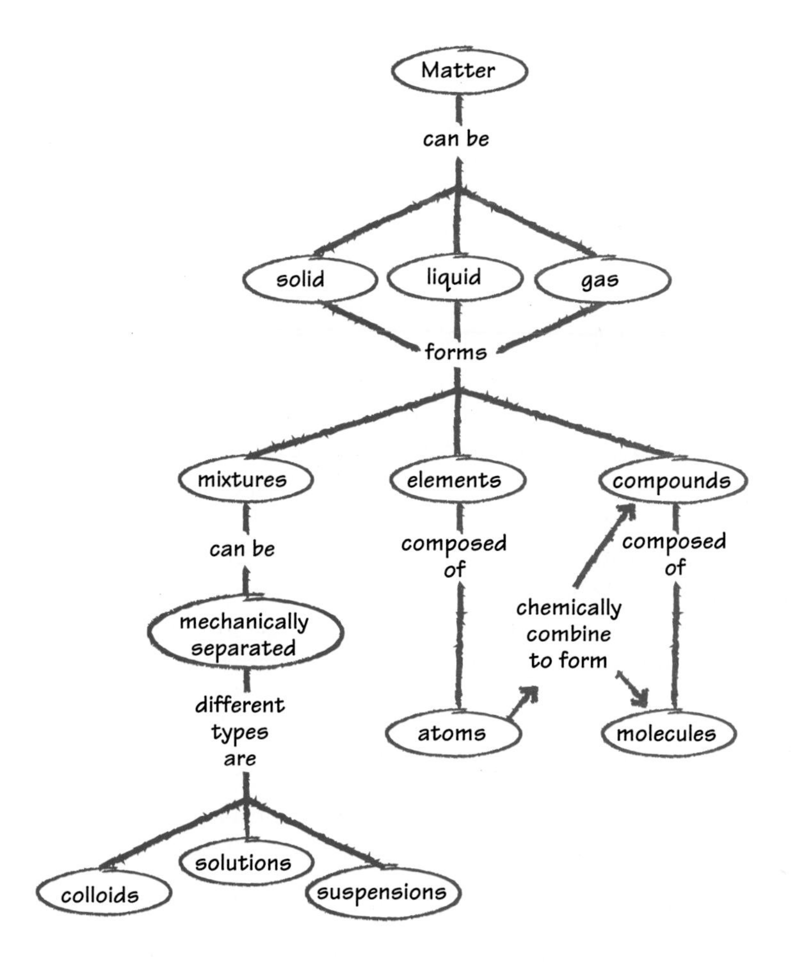

Evaluating Concept Maps

Again, there is no single correct concept map. However, consider the following criteria as you evaluate your students' concept maps.

- how comprehensive the map is (Are all relationships shown?)
- how clearly the concepts are linked (Are proper relationships between concepts shown? Are linkage terms used between all concepts?)
- overall clarity of presentation (Could the map be simpler? Is it redundant? Is it logically arranged? Are linkage terms used properly?)

Used properly, concept maps can increase comprehension, improve retention, and sharpen your students' study skills. Concept maps are a valuable addition to any student's arsenal of learning strategies.

Concept maps are a valuable addition to any student's arsenal of learning strategies.

Communicating Science

One of the most important skills that students can acquire is the ability to communicate what they have learned, both orally and in writing. Students' comprehension is enhanced when they are called upon to reformulate in their own words what they have learned. You can help students develop and communicate their mastery of new ideas in novel ways—for example, by writing for one another or for some audience other than the teacher. Students' comprehension improves when they are called upon to reformulate in their own words what they have learned.

By engaging their verbal skills, you can help students develop and communicate their mastery in novel ways.

You may ask your students to communicate what they have learned in many different ways. For example, assignments may require students to interpret a passage, illustrate a paragraph, write a headline, label a diagram, or write a caption for a photo or drawing. These strategies can enhance students' understanding by employing their multiple intelligences.

Reading and Writing in the Classroom

The time spent in helping students prepare to read is critical in fostering comprehension. Two strategies are particularly important in the pre-reading phase of instruction: building on prior knowledge and establishing a purpose for reading.

The Power of Prior Knowledge

The amount of prior knowledge that students have about a topic directly influences their comprehension of that topic. The more students know about something, the easier it is for them to grasp new information about it. Helping students identify the information they already know about a topic assists them in relating the new information to existing knowledge.

Students tend to retain misconceptions they may have about a topic. If the text information seems to conflict with their preconceptions, students may ignore or reject new information. It is therefore extremely important to identify these misconceptions so that they may be dispelled.

Reading to Understand

One way to establish a purpose for reading is to provide a study guide with questions that students can answer as they read. You can also help students learn to make their own study guides. First teach students to preview a chapter or unit by looking at all the headings, illustrative material, and terms and phrases in boldface or italics. This technique helps students gain a feel for the main concepts and helps them build a basic structure for the new information. Then show them how to use the captions, headings, and highlighted words to devise study guide questions to answer after reading the material. Following this method, students will read with a purpose in mind—a purpose of their own devising. You may also wish to consider using the text on audiocassette, which is an excellent means for previewing material with students.

Writing to Understand

Studies show that writing is an effective tool for improving reading. As students write, they are creating a text for others to read. The most important advice to give students about scientific writing is to strive for clarity and accuracy. These characteristics can often be achieved with simple vocabulary and short sentences. One useful approach might be to have students imagine that they are writing for a younger audience.

Teacher to Teacher

Find out what your colleagues have to say about this hot topic in their article "Teaching and Learning Science Through Writing" on page 100.

Process Skills

Process skills are a means for learning and are essential to the conduct of science. Perhaps the best way to teach process skills is to let students carry out scientific investigations and then to point out the process skills they used in the course of the investigations. Look for and encourage the use of the following skills in your classroom:

Observing

An observation is simply a record of a sensory experience. Observations are made using all five senses. Scientists use observation skills in collecting data.

Communicating

Communicating is the process of sharing information with others. Communication can take many different forms: oral, written, nonverbal, or symbolic. Communication is essential in science, given its collaborative nature.

Measuring

Measuring is the process of making observations that can be stated in numerical terms. All scientific measurements should be given in SI units.

Comparing

Comparing involves assessing different objects, events, or outcomes for similarities. This skill allows students to recognize any commonality that exists between seemingly different situations. A companion skill to comparing is contrasting, in which objects, events, or outcomes are evaluated according to their differences.

Contrasting

Contrasting involves evaluating the ways in which objects, events, or outcomes are different. Contrasting is a way of finding subtle differences between otherwise similar objects, events, or outcomes.

Organizing

Organizing is the process of arranging data into a logical order so the information is easier to analyze and understand. The organizing process includes sequencing, grouping, and classifying data by making tables and charts, plotting graphs, and labeling diagrams.

Classifying

Classifying involves grouping items into like categories. Items can be classified at many different levels, from the very general to the very specific.

Analyzing

The ability to analyze is critical in science. Students use analysis to determine relationships between events, to identify the separate components of a system, to diagnose causes, and to determine the reliability of data.

Inferring

Inferring is the process of drawing conclusions based on reasoning or past experience.

Hypothesizing

Hypothesizing is the process of developing testable explanations for phenomena. Testing either supports a hypothesis or refutes it.

Predicting

Predicting is the process of stating in advance the expected result of a tested hypothesis. A prediction that is accurate tends to support the hypothesis.

Cooperative Learning

Cooperative learning is a teaching technique that brings students together to learn in small, heterogeneous groups. In these groups, students work interdependently without constant and direct supervision from the teacher. Assignments are structured so that everyone contributes. Challenges as well as rewards are shared. Brainstorming, lively discussion, and collaboration are the hallmarks of the cooperative-learning classroom.

What It's Not!

- Cooperative learning is not the same as ability grouping, where a teacher divides up the class in order to instruct students with similar skills.
- Cooperative learning is not having students sit side by side at the same table to talk while they complete individual assignments.
- Cooperative learning is not assigning a task to a group in which one student does the work and the others get equal credit.

Benefits of Cooperative Learning

- **Cooperative learning models the scientific experience.** Students working in groups learn about the joys as well as the frustrations involved in scientific inquiry. Cooperative learning models real scientific experience in which scientists work together, not in isolation, to solve difficult problems. With cooperative learning, the classroom becomes a fertile environment for ideas and novel solutions.
- **Cooperative learning empowers and involves students.** Cooperative learning raises students' self-esteem because they are learning something on their own through cooperation, rather than being handed prepackaged knowledge. It helps students become self-sufficient, self-directed, lifelong learners. In a cooperative learning environment, students are less dependent on you for knowledge.

Teacher to Teacher

Find out what your colleagues have to say about this hot topic in their article "Meeting the Needs of the Academically Gifted" on page 78.

As students work toward a common goal, they come to recognize commonalities that cut across differences related to ethnicity, socioeconomic background, and gender.

- **Cooperative learning serves the heterogeneous classroom.** With group work, everyone has the chance to participate, and everyone has a role to play. As students join forces to achieve a common goal, they come to recognize commonalities that cut across differences related to ethnicity, socioeconomic background, and gender. Likewise, cooperative learning provides an excellent vehicle for students of differing ability levels to work together in a positive way. Challenged students can interact successfully with average and advanced students and in so doing can learn that they too have something to offer.

Teacher to Teacher

Find out what your colleagues have to say about this hot topic in their article "Ensuring Girls' Success in Science" on page 66.

- **Cooperative learning strengthens interpersonal skills.** Group tasks are structured so that students must cooperate to succeed. Students quickly understand that they will "sink or swim" together by how constructively they interact. Consequently, students develop important interpersonal and social skills that help them function in a group setting and that will ultimately benefit them socially, at work, and in other situations.
- **Cooperative learning develops appropriate social skills.** When doing cooperative group work, students channel their energies into constructive tasks while satisfying their fundamental need for social interaction.
- **Cooperative learning is an effective management tool.** Establishing cooperative learning in the classroom requires you to relinquish some control, so the students themselves can become responsible for building their own knowledge. Working in groups to probe and investigate ideas, answer questions, and draw conclusions about observations allows students to discover and discuss concepts in their own language. When students learn through cooperation, the knowledge derived becomes their own, not just a loan of your ideas or those from the textbook.
- **Cooperative learning increases achievement.** Since the 1920s, there has been extensive research on cooperative learning techniques. Results clearly indicate that cooperative learning promotes higher achievement for all grade levels in all subject areas.[1]

Using Cooperative Learning: The Basics

Group Size

Although group size will vary depending on the activity, the optimum size for cooperative learning is between three and four students. For students unaccustomed to this learning style, keep the group size to about two or three students.

Group Goal

Students need to understand what is expected of them. Identify the group goal, whether it be to master specific objectives or to create a product such as a chart, a report, or an illustration. Identify and explain the specific cooperative skills required for each activity.

---

[1] Johnson, Johnson, Houlbec, and Roy. *Circles of Learning, Cooperation in the Classroom.* Association for Supervision and Curriculum Development. © 1984

There is no single set of cooperative learning strategies that will work with all students in all situations. However, the following strategies may provide you with insight and guidance in developing your own set of strategies that will work for your students.

Balance the needs of students of all levels and learning styles

Your ultimate goal will be to ensure that all students are able to work effectively in any group. Initially, however, you may wish to develop special grouping strategies to foster the growth of learners having difficulty and second-language learners and to assure gifted students that their grade will not be affected by slower learners. More information is provided about grouping strategies on the next page.

Positive Interdependence

A learning activity becomes cooperative only when everyone realizes that no group member can be successful unless all group members are successful. The "we're all in this together" part of group work is the positive interdependence. Encourage positive interdependence by assigning each student some meaningful role or allow students to do this themselves. You can also encourage positive interdependence by dividing materials, resources, or information among group members.

Individual Accountability

Each group member should have some specific responsibility that contributes to the learning of all group members. At the same time, each group member should reach a certain minimum level of mastery.

Meeting Individual Needs with Cooperative Learning

Cooperative learning is an effective tool for meeting the individual needs of your students. Cooperative learning builds relationships among students where relationships might not have developed before. Students are required to interact with each other as individuals with common goals. In so doing, students learn more about each other's personal characteristics, and as a result, many stereotypes are destroyed.

Have a clearly defined goal

When you tell students what is expected of them, be sensitive to their special needs. Be sure that each student understands the group goal and his or her own personal responsibility.

It is also important that assignments be specifically appropriate for groups. In other words, simply having students fill in the blanks of a worksheet or answer the end-of-unit questions is not creating an adequate cooperative learning assignment. Students need tasks that cannot easily be completed alone. Students should see that if they work together, the end product will be better and more complete than if they had worked alone.

Answer questions only when the whole group has the same question. It's a good idea to designate one person per group as the liaison between you and the group.

Praise success

If students seem unmotivated or feel that their individual tasks are unimportant to the success of the group, you may wish to consider offering group rewards. Reward the groups as they successfully complete each activity. Reward successful project results as well as positive interaction and effective group process skills. However, rewards should not become automatic. They should be used only for the short term. For the long term, students should take pride in their group's achievements and should benefit from the knowledge that these achievements contribute to their success as individuals.

Encourage interpersonal problem solving within groups

Pulling a disruptive student out of a group is sometimes necessary, but be sure that the isolation is only temporary. Difficult students need the support of others. Build into each group the spirit of encouraging each other. Suggest that groups evaluate their own performance after an activity is finished. Encourage students to suggest solutions to problems without criticizing individuals.

Teacher to Teacher

Find out what your colleagues have to say about this hot topic in their article "Yes, Teaching Students to Argue Is a Good Idea . . . No, I Am Not Crazy!" on page 44.

Grouping Strategies

With cooperative learning, you can either place students in particular groups or assign students to groups at random. There are advantages to both approaches. At first, however, it is recommended that you assign students to particular groups.

The *Assessment Checklists & Rubrics* booklet and the *One-Stop Planner CD-ROM* contain several checklists and other tools to help you implement cooperative-learning groups.

Assigned Grouping

Composing groups yourself lets you create groups that are heterogeneous in terms of academic ability, gender, ethnicity, and cultural background. Heterogeneous groups are preferred because cooperation among diverse students not only teaches the widest range of interpersonal skills but also promotes frequent exchange of explanations and greater perspective in discussions. This increases depth of understanding and retention of concepts.

To create effective heterogeneous groups, balance each group with students who have different strengths. First decide who your resource students are. These are students you think will facilitate group work—either because of their academic ability or because of their interpersonal skills. Assign at least one resource student to each group. Distribute students who may be disruptive and students who lack academic skills evenly throughout the groups. Avoid putting close friends together to prevent cliques from disrupting teamwork. Put students who have limited English proficiency in groups with bilingual students who can act as translators.

Random Grouping

Random grouping can be especially effective with experienced cooperative learners or if you plan to change group membership often. To create random groups, you can simply have students count off from one to five. All of the ones form a group, all of the twos form another group, and so on.

There are many other fun ways to assign groups randomly. For example, you can hold a lottery in which students pick numbers out of a hat. Numbers one to four form one group, numbers five to eight form another; and so forth. You can also use cards naming sets of a particular type of item. All students who draw items belonging to the same set form a group. For example, all students whose cards name types of flowers belong in one group, students whose cards name farm animals belong in another group, those with cards naming heavy-metal bands form a third group, etc. Students have a fun and lively time discovering who belongs in the same group.

You can also combine lesson content with assigning groups. First decide how many students you want in each group. For each group, write a different scientific term or principle on a flashcard. Then for each group's term, list on separate cards the definition of the term, a synonym, or an example of what the term means. Mix up the cards and hand them out as students come into the room or once students are seated. Students use the cards as clues to find the others in their group.

Assigning Roles

Assign roles at first; students can choose their own later. Assigning roles is very important, especially at first. Consider the behavior patterns of students, and assign roles that will complement those patterns. Group work needs to be structured so that everyone has a part to play. In other words, there needs to be positive interdependence. Just as members of a surgical team work together, with each person contributing his or her own special skill, students work effectively in teams when everyone has a unique role that is vital to the group's success.

Some examples of useful roles are listed here. Use as many as you need, modify them, combine them, or invent roles yourself.

- **Facilitator** The facilitator is a leadership role. The facilitator keeps an activity running smoothly by presiding over the work flow. He or she manages the group so that all members have a chance to talk, questions are answered, students listen to one another's ideas, and ideas are substantiated with reasons and explanations.
- **Recorder** The recorder records data and answers questions posed to the group.
- **Reporter** The reporter explains the group's findings to the teacher or the entire class.
- **Safety officer** The safety officer makes sure safety practices are followed and notifies the teacher of any unsafe situations.
- **Checker** The checker makes sure that everyone has finished his or her worksheet or other individual assignment.
- **Materials manager** The materials manager gathers activity materials at the outset, monitors their use during the activity, and organizes the cleanup and return of materials to their proper place after an activity.

Again, assign roles carefully, especially at first, taking into account students' behavior patterns. A shy student might be most comfortable as a recorder, while a student who likes attention might make the best reporter. The facilitator is a role that some students will always want and others will avoid. Be careful not to stereotype. Sometimes the most unlikely students will make the best leaders.

Assessing Cooperative Learning

Assessment within a cooperative-learning setting is not as difficult as it may seem. Like any other assessment, you must determine in advance what you would like to assess and to what degree. You will also need to develop some slightly different monitoring skills.

A variety of checklists designed for monitoring and assessing cooperative group work are provided in the *Assessment Checklists & Rubrics* booklet and on the *One-Stop Planner CD-ROM*. Both student and teacher checklists are provided.

Monitoring Groups

Resist the temptation to get caught up on paperwork as the groups do their work—this is the time to observe, monitor, and coach. As you monitor the groups, you can reinforce cooperative behavior with a formal observation sheet. Record how many times you observe each student using a collaborative skill, such as contributing ideas or asking questions.

If a group seems hopelessly confused or "stuck," you can intervene to guide students to a solution. But make sure students have the opportunity to reason through problems themselves first. Consider the following differences between direct supervision and the kind of monitoring that supports cooperative learning.

Direct Supervision	Supportive Monitoring
Lecturing	Giving feedback
Disciplining	Encouraging problem solving
Telling students what to do	Providing resources
Leading discussions	Observing

What to Assess

What should you assess in a cooperative learning activity? Individual success? Group success? Cooperative skills? Actually, many teachers find it useful to evaluate all three. And there are many ways to assess each of these areas.

Individual success can be evaluated by asking students to fill out answers to a worksheet as they progress through an activity; by having them record, analyze, and submit data; or by having them take a quiz. Some activities are structured so that each student turns in a product, such as a report or a poster, that can be individually graded.

Group success is evaluated according to how well the group accomplished its assigned task. Was the task completed? Were the results accurate? If not, were errors explained and accounted for? Criteria such as these provide a framework for group evaluation.

Cooperative skills are evaluated based on your observations of students' behavior in their group. Evaluating students' use of cooperative skills will motivate students to use them. If you intend to grade cooperative skills, it is helpful to use a formal observation checklist as you monitor students at work. Log the frequency with which group members exhibit cooperative skills or disruptive behavior.

If you wish to compute a single overall grade, assign a weight to each of the three grade components, stressing the factors that you consider most important. Use cooperative learning to meet the needs of your students, and enjoy!

Critical Thinking

Critical thinking skills are essential for making sense of large amounts of information. Too often, science lessons leave students with a set of facts and little ability to integrate those facts into a comprehensible whole. Requiring students to think critically as they learn improves their comprehension and increases their motivation.

Loosely defined, critical thinking is the ability to make sense of new information based on a set of criteria. Critical thinking skills draw on higher-order thinking processes, especially synthesis and evaluation skills. Critical thinking takes a number of different forms, a few of which are described below.

Validating Facts

This type of critical thinking involves judging the validity of information presented as fact. Too often, people will accept as valid almost any statement, no matter how outrageous, as long as it comes from a supposedly authoritative source. It is important for scientists to treat all untested data with suspicion, no matter how reasonable it may seem. Students may validate facts in a number of ways: by observing, by testing, or by rigorously examining the logic of the so-called fact.

Making Generalizations

A scientist must often be able to identify similarities among disparate events. Generalizations are drawn based on a limited set of observations that can be applied to an entire class of phenomena. For example, one does not have to test every known substance to make the generalization that solid substances melt when heated. Generalizations allow scientists to make predictions. Once the rule is known, future outcomes can be forecast with a high degree of confidence.

It is important that students base their generalizations on an adequate amount of information. A generalization that is formed too quickly may be wrong or incomplete or may lead the student down a dead-end path.

Making Decisions

Many students would not regard science as a field requiring decision-making skills. But in fact, scientists must make decisions routinely in the course of their work. Any time a scientist works through a problem or develops a model, a whole series of decisions must be made. A single faulty decision can throw the entire process into disarray. Making informed decisions requires knowledge, experience, and good judgment.

Interpreting Information

Having all the information in the world is useless unless one also has the tools to interpret that information. Scientists must know how to separate the meaningful information from the "noise." Information can come in any form detectable by the five senses. It is important that scientists and students alike interpret information to determine its meaning, validity, and usefulness.

Requiring students to think critically as they learn improves their comprehension and increases their motivation.

Environmental Awareness

No species affects its surroundings as dramatically as does the human species. Because of the publicity given to events and processes such as the Chernobyl incident, the destruction of the rain forests, and the depletion of the ozone layer, people have come to realize the global impact that human actions can have. It is incumbent on the educational system to promote environmental awareness among students. To promote an appreciation of the environment among students, it is important to present environmental issues in a way that students can easily grasp.

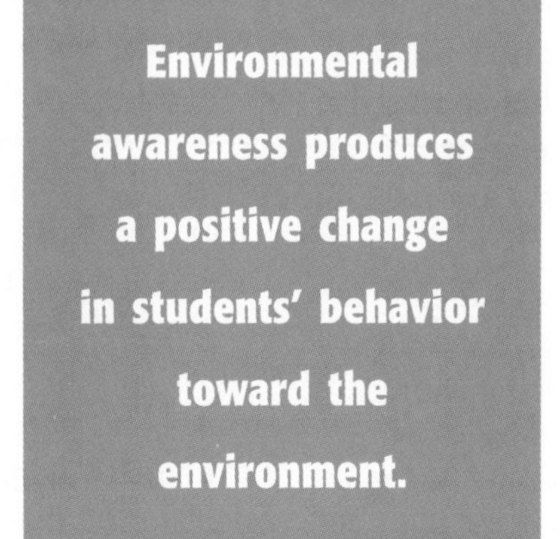

Environmental issues run the gamut from local to global. While large-scale problems get headlines, they can be hard to grasp for many students who may never have observed those problems directly. In most cases it is best to start building students' awareness by introducing them to local issues. Local issues not only are more relevant to their lives but also are more likely to lead to direct involvement.

Environmental awareness serves two purposes: it promotes understanding of the living world and the place of humans within it, and it produces a positive change in students' behavior toward the environment. This program pursues both goals and can help you draw students' attention to these issues.

You may involve students directly in environmental issues by using the suggested activities in the Pupil's Edition. The Environment Connection box, such as the one shown below, contains relevant environmental information that you can use to add depth to the topic at hand and to encourage discussion or some other action. In addition, the Annotated Teacher's Edition contains a number of Environment Connection strategies that can also encourage discussion.

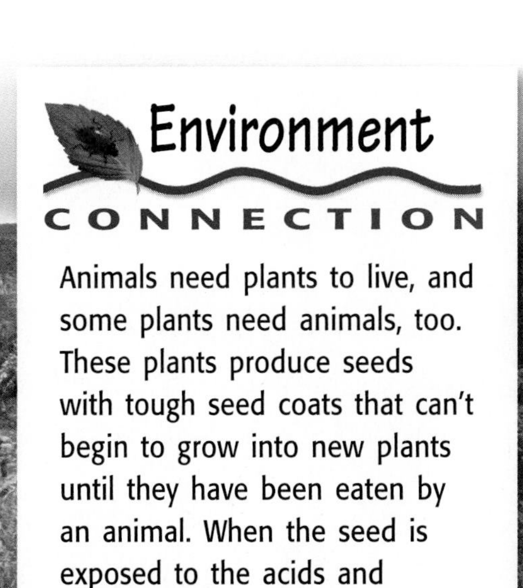

Multicultural Instruction

The success of our nation would not be possible without the contributions of the many cultures and ethnic groups that make up our country. Multicultural instruction serves to ensure that all students have the chance to learn, to succeed, and to become whoever they would like to become, regardless of race, gender, socioeconomic background, or disability. Multicultural instruction affirms the positive nature of this country's diversity by helping students develop an open mind, a positive self-concept, and a realistic understanding of the world that surrounds them.

Meeting the needs of culturally diverse students is perhaps the most demanding challenge faced by today's teachers. We must constantly strive to arouse adolescent curiosity, minimize risks of failure, and be as responsive as possible to our individual students' needs. The more culturally relevant we make our science programs, the better we will be at serving our changing class populations. These challenges are especially difficult with middle-school students because they are also going through the physical and emotional changes associated with adolescence.

By their very nature, middle-school science programs have the potential for promoting the full development of individual learners, especially when science classrooms are perceived as places of inquiry and discovery. When such environments exist in science classrooms, students become more successful learners.

Multicultural instruction affirms the positive nature of this country's diversity by helping students develop an open mind, a positive self-concept, and a realistic understanding of the world that surrounds them.

Relevance, Positive Self-Concept, and Multiculturalism

Let your students help you incorporate multicultural learning into the classroom by allowing them the freedom to express their feelings and attitudes during your classes. With a program tailored specifically to the personal experiences of your students, you will find that your students are more curious about the world around them. With an increased level of relevance, learning is more important to young thinkers.

Likewise, it is very important to be sensitive to the cultural identities of your students. As a teacher, you have the opportunity to model behavior demonstrating that all cultures have equal value. Showing a healthy respect for your students' cultures will help each of your students achieve and maintain a positive self-concept.

Using Multicultural Instruction

A strong program of multicultural instruction can begin by implementing a few basic strategies. While none of the strategies are exclusively multicultural, they can provide proper contexts and situations that capitalize on the cultural backgrounds of students.

- Recognize and convey to students that all languages are equally valid. Explain, however, that learning English increases the range of opportunities available to the individual.

- Draw special attention to the diversity of role models in the textbook. At every opportunity, provide information about past and current scientists from diverse cultures.

  The power of such role models should not be underestimated. Role models may create interest and motivation, and may even influence a student's pursuit of a career.

- Use cooperative learning to diversify student groups. You will find that students develop more of an open-minded awareness as well as more positive, accepting, and supportive relationships with peers. Labels concerning ethnicity, gender, learning ability, social class, and disability cease to exist.

- Peer and cross-age tutoring is an excellent strategy for fostering better understanding among individuals. Peer tutoring involves students tutoring students their own age. Cross-age tutoring involves older students tutoring younger students. These strategies are beneficial to both the tutors and the students being tutored. In using either of these strategies, be very careful when pairing students. Although this is an excellent opportunity to integrate students, both the tutor and the student must be willing participants.

- Take every opportunity to relate science to personal experiences. Invite your students to discuss any of their own experiences that may apply. You may discover some very relevant connections and analogies, and the learning process will become more interactive and personalized to the class. This process might also add to the richness of the class by highlighting the cultural differences among your students.

- Allow students to select independent projects that are relevant to their own world. These projects should permit students to create new, positive avenues of self-expression from their own experiences. Students find opportunities to select, design, and articulate their own interest within science programs while developing their creative thinking and problem-solving skills. In addition, these activities promote the development of positive attitudes toward general academics, social interactions, and the study of science.

Supporting Multicultural Instruction with the Text

Science is for everyone, and this program is designed to serve the multiethnic and multicultural classrooms of today. Students, regardless of their ethnic backgrounds, will not have to look hard to find positive role models. In addition, content that shows events, concepts, and issues from diverse ethnic and cultural perspectives is provided. As students work through this program they will come to understand that science is a human endeavor that has been advanced by the contributions of many cultures and ethnic groups.

To add depth to your multicultural instruction, the wraparound margins of the Annotated Teacher's Edition periodically include a feature called Multicultural Connection. This information provides activities to help you focus on cultural diversity, highlighting the individuality and contributions of different ethnic groups.

Meeting Individual Needs

Obviously, to teach effectively you must be able to reach every individual in your class. This is seldom easy, given the diverse nature of most of today's classrooms. In addition, certain students present special challenges. Dealing adequately with these students requires special preparation and strategies. In many cases a minimal amount of preparation is sufficient to make the classroom a place where all can learn. Some of the more common situations you are likely to encounter are discussed below.

Learners Having Difficulty

Learners having difficulty are those who, for any number of reasons, are liable to perform poorly and who have a high probability of dropping out of school. This program is engaging and interesting throughout, appealing to all students. Throughout, clear easy-to-read prose and straightforward, attractive graphics reduce the potential for students to become bored. The style of this program is intentionally friendly and unintimidating.

Additional activities and teaching suggestions for learners having difficulty are provided under the Meeting Individual Needs: Learners Having Difficulty heading in the wraparound margins of the Annotated Teacher's Edition.

Additional strategies for meeting the needs of learners having difficulty can be found in two articles in this booklet. In the article "Motivate the Unmotivated with Scientific Discrepant Events," on page 60, an experienced teacher shares his technique for piquing students' interest in science. If behavioral problems are the cause of students' difficulty, you may find reading "Strategies for Improving Student Behavior," on page 44, to be helpful.

Second-Language Learners

This program is ideal for students who are not proficient in English. Specific suggestions for second-language learners are found in the wraparound margins of the Annotated Teacher's Edition. Look for the heading Meeting Individual Needs: Learners Having Difficulty. Also, when you are looking for strategies to involve second-language learners in lessons, look for activities that have a sheltered English label.

Additional strategies for second-language learners or other students struggling with the text include Reading Strategies that can also be found in the Annotated Teacher's Edition. With these strategies you can guide students through the lesson using mnemonics, prediction guides, or in-class activities. Also look for the Reteaching heading that suggests another way to present an important concept in a lesson.

Naturally, students for whom English is a second language have special educational needs. To address their needs and to effectively teach learners with limited English proficiency requires a basic knowledge of relevant issues. The article on page 72 of this booklet, "Teaching Science to Students with Limited English Proficiency," lends insight into these issues as well as provides a method for planning lessons that effectively address the issues.

Meeting Individual Needs

Learners Having Difficulty

Assemble a group of objects for students to investigate, observe, and take apart if they wish. Possible objects include a flashlight, a pen, and a stapler. After students examine each object, ask them to speculate about how each object works. To get them started, suggest that they make a list or draw a diagram of the different parts in each object. Then they can proceed by writing down the possible function of each part. Ask students to summarize the ways in which they behaved like scientists during this activity.

Gifted Learners

The difficulty of teaching gifted students lies in keeping them interested, motivated, and challenged. Gifted students who are inadequately challenged may become bored, withdrawn, or even openly disruptive. This program includes many activities suitable for even the most advanced student. Open-ended activities, in particular, are especially suited for gifted students.

Additional activities and teaching suggestions for gifted students are found in the wraparound margins of the Annotated Teacher's Edition. Look for the heading Meeting Individual Needs: Advanced Learners.

This program emphasizes creative problem solving. In many cases there is no single right answer to a problem or question, so students' answers can reflect their individual abilities. This approach is ideal for gifted students as they may extend the activities to fit their interests and talents. For example, the following strategy for advanced learners comes from an Annotated Teacher's Edition in this series. A fresh perspective on this topic and additional strategies for teaching gifted students are explained in "Meeting the Needs of the Academically Gifted," on page 78.

Meeting Individual Needs

Advanced Learners Encourage interested students to investigate Darwin's voyage and similar long-distance travel by explorers in the 1800s in greater detail. Topics for reports include the types of ships used for travel in that era, the kinds of food eaten by explorers, and the sophistication and thoroughness of maps in the 1800s.

Physically Impaired Students

Are all parts of your classroom accessible to all students? Adapt the classroom to enable physically impaired students to engage in the same activities as other students. Encourage your students to assist physically impaired students. Make your classroom as easy to move about in as possible. Remove or bypass any obvious barriers. If the student uses a wheelchair, make the aisles wide enough to accommodate the chair. Make sure that the student can reach any equipment he or she needs.

As much as possible, adapt the classroom to make it possible for physically impaired students to engage in the same activities as other students. Use a mobile demonstration table so that it can be moved to different areas of the room for maximum visibility.

Visually Impaired Students

Seat students with marginal vision near the front of the room to maximize their view of both you and the chalkboard, or assign a student to make copies of what you write. You could also assign a neighboring student to quietly explain all visual materials in detail as they are presented.

Students who are completely blind should be allowed to become familiar with the classroom layout before the first class begins. Promptly inform these students of any changes to your classroom layout. Whenever possible, provide Braille or taped versions of all printed materials. Your school librarian may be able to help procure these items. Students who are blind may also use handheld devices for converting written text into speech.

Speech-Impaired Students

Mainstreaming speech-impaired students is generally not very difficult. Patience is essential when dealing with speech-impaired students, however. For example, resist the temptation to finish sentences for a student who stutters. Also pay attention to nonverbal cues, such as facial expression and body language.

Be supportive and encouraging. You need not leave the speech-impaired student out of normal classroom discussions. For example, you may call on a speech-impaired student to answer a question and then allow the student to write out his or her response on the chalkboard or overhead projector. Use multisensory materials whenever possible to create a more comfortable learning environment for the speech-impaired student.

Hearing-Impaired Students

If you have students with hearing-impairments in your class, remember to always face the class while speaking. Minimize classroom noise, and arrange seating in a circle or semicircle so that students can see one another. This arrangement facilitates speech reading. Speak in simple, direct language and avoid digressions or sudden changes in topic. During class discussions, periodically summarize what students are saying and repeat students' questions before answering them. Use visual media such as filmstrips, overhead projectors, and close-captioned films when appropriate. You might arrange a buddy system in which another student provides copies of notes about activities and assignments.

A student who is completely deaf may require a sign-language interpreter. If so, let the student and the interpreter determine the most convenient seating arrangement. When asking the student a question, be sure to look at the student, not at the interpreter. If the student also has a speech impairment, group assignments for oral reports may be advisable.

Learning-Disabled Students

Learning disabilities are any disorders that obstruct a person's listening, reasoning, communication, or mathematical abilities, and they range from mild to severe. An estimated 2 percent of all adolescents have some type of learning disability. Learning disabilities are the most common type of disability. To help learning-disabled students succeed, provide a supportive and structured environment in which rules and assignments are clearly stated. Use familiar words and short, simple sentences. Repeat or rephrase your instructions as needed.

Students may require extra time to complete exams or assignments, with the amount of extra time being dependent on the severity of their disability. Some students may need to tape-record lectures and answers to exam questions. For those who have difficulty organizing materials, you might provide chapter or lecture outlines for them to fill in. Having peer tutors work with learning-disabled students on specific assignments and review materials can be effective.

Computer-assisted instruction is an extremely useful tool for some learning-disabled students. This mode of instruction can even help these students develop good learning skills. For learning-disabled students, computers serve as a tireless instructor with unlimited patience. In addition, students receive simplified directions; proceed in small, manageable steps; and receive immediate reinforcement and feedback with computerized instruction.

Students with Behavioral Disorders

Behavioral disorders are emotional or behavioral disturbances that hinder a student's overall functioning. The behaviorally impaired may exhibit any of a variety of behaviors, ranging from extreme aggression to complete passivity.

Obviously, no single teaching strategy can accommodate all behavioral disorders. In addition, behavioral psychologists disagree on the best way to deal with students who have behavioral disorders. As a general rule, try to be fair and consistent yet flexible in your dealings with behaviorally disabled students. Make sure to state rules and expectations clearly. Reinforce desirable behavior or even approximations of such behavior, and ignore or mildly admonish undesirable behavior.

Because learning disabilities often accompany behavioral disorders, you might also wish to refer to the guidelines for learning disabilities.

Materials and Equipment

This program is designed to be teachable even by those with a limited budget for materials. Most activities use common household items that can be brought to class by students or parents or that can otherwise be easily obtained. The *One-Stop Planner CD-ROM* contains a letter that you can send to parents to request materials donations. This letter makes it easy for you to invite donations from parents to keep your expenses as low as possible.

As you can see from the sample shown below, *Holt Science and Technology* is designed around readily available materials and equipment. The amounts shown in the listing are for one group or student. More specific materials and information can be found with the lab or investigation in the Annotated Teacher's Edition.

For a comprehensive listing of the materials you would need in order to teach all of the labs and investigations in *Holt Science and Technology*, please refer to the Master Materials List in each book.

MATERIALS AND EQUIPMENT		QuickLab	Investigate!	LabBook
CONSUMABLE	**AMOUNT***	PAGE NO.	PAGE NO.	PAGE NO.
Aluminum foil, approx. 5 × 5 cm	1			586
Aluminum, approx. 5 × 1 cm	6			580
Antacid tablet	1	404		572
Bag, paper lunch	1		35	
Bag, sealable plastic	1		373	
Bag, sealable plastic	6	381		524
Baking powder	1 tsp		373	
Baking soda	1 tsp	381		
Baking soda	1 tbsp		213	
Baking soda	4 g	87		526
Balloon, assorted colors, round	1			543
Balloon, long, 12 in.	1			568
Battery, 6 V	1	290, 291		592
Battery, D-cell	2			592
Battery, D-cell, weak	1		351	
Borax	approx. 2 tbsp			555
Bottle cap	4–6			582
Cabbage, red	1 leaf		373	
Calcium chloride	2 tsp			530
Calcium chloride solution	approx. 10 mL			594
Cal...				

Science Kit

For teachers who would rather have the convenience of purchasing supplies through the mail, Science Kit® is the official materials and equipment supplier for this program. For your convenience, Science Kit® offers kits that contain the materials and equipment you will need to teach each unit of this program. Science Kit® also provides materials-ordering software on CD-ROM designed specifically for *Holt Science and Technology*. This software allows you to create an electronic materials list, complete with item number. Using this software, you can order complete kits or individual items, quickly and efficiently. Or if you prefer, you can order needed materials and equipment individually as necessary. For more information about this software, contact your HRW representative, call Science Kit® directly at 1-800-828-7777, or visit the Web site: www.sciencekit.com.

Assessing Student Performance

A Comprehensive Approach to Assessment

Developing strategies for assessing student progress is an important step in realizing the goals you have for your students. Students pay the most attention to those aspects of a lesson on which they know they will be graded. Teachers who want their students to be successful should therefore teach with continual assessment in mind.

You will find that most of the tests and assessment activities in this program are designed to *teach* as well as to evaluate comprehension and performance. This emphasis can help correct the preoccupation with measuring and sorting students. The suggestions here are intended to aid you in your primary task: teaching.

Assessment Aids

Holt Science and Technology includes a wide variety of assessment aids to help you measure your students' mastery of the concepts and processes covered. Assessment materials contained in the Pupil's Edition include Review questions at the end of each section and a Chapter Review at the end of each chapter. The Annotated Teacher's Edition helps you assess students' comprehension of sections by providing two quiz questions and alternative assessment options at the end of each section. The following materials are also available for comprehensive and convenient assessment.

Chapter Tests with Performance-Based Assessment

This booklet is designed to help you determine how well students have mastered the content and scientific methods. There is a Chapter Test and a Performance-Based Assessment for each chapter of the Pupil's Edition. The Chapter Tests are four pages in length and include questions that range from those that prompt students to recall specific content to those that require them to use higher-order thinking skills.

Performance-Based Assessments (PBA) allow you to evaluate students' abilities to solve problems using the tools, equipment, and techniques of science. A Teacher's Preparatory Guide at the beginning of each PBA provides you with helpful information. Each PBA also includes a specific rubric to clarify expectations and simplify grading.

Assessment Checklists & Rubrics

This booklet contains over 40 different checklists for student self-evaluation, peer evaluation, and teacher assessment. Checklists are also provided to aid ongoing assessment.

You will also find in this booklet a variety of assessment rubrics that serve as models for grading writing assignments, portfolios, reports, presentations, experiments, and technology projects.

Progress reports are also available to help you keep track of your students' progress over time.

Most of the tests and assessment activities in this program are designed to teach as well as to evaluate comprehension and performance.

One-Stop Planner CD-ROM with Test Generator

The *One-Stop Planner CD-ROM* for Macintosh® and Windows® includes a *Test Generator* for your convenience in building assessment tools. The *Test Generator* contains Chapter Tests and Performance-Based Assessments for each chapter of the Pupil's Edition. The *Test Generator* is easy to use, includes graphics, and is fully customizable. With this tool you can easily change questions or add questions of your own.

The *One-Stop Planner CD-ROM* also contains the *Assessment Checklists & Rubrics* booklet described above so that you may customize checklists, rubrics, and progress reports to meet your specific needs.

Test Generator: Test Item Listing booklet

The *Test Item Listing* booklet gives you quick reference to all of the items stored on the *Test Generator,* which is located on the *One-Stop Planner CD-ROM.* The booklet is a handy way to preview the tests and questions contained in the *Test Generator.* The booklet also contains the answers to section review questions, chapter review questions, and chapter tests, as well as all Performance-Based Assessments and their corresponding teaching strategies and answers.

Chapter Organizer

CHAPTER ORGANIZATION	TIME MINUTES	OBJECTIVES	LABS, INVESTIGATIONS, AND DEMONSTRATIONS
Chapter Opener pp. 4–5	45		**Investigate!** Mission Impossible? p. 5
Section 1 **Exploring Physical Science**	90	▶ Describe physical science as the study of energy and matter. ▶ Explain the role of physical science in the world around you. ▶ Name some careers that rely on physical science.	
Section 2 **Using the Scientific Method**	90	▶ Identify the steps used in the scientific method. ▶ Give examples of technology. ▶ Explain how the scientific method is used to answer questions and solve problems. ▶ Describe how our knowledge of science changes over time.	**Demonstration,** p. 12 in ATE **QuickLab,** That's Swingin'! p. 15 **Discovery Lab,** Exploring the Unseen, p. 626 **Datasheets for LabBook,** Exploring the Unseen, Datasheet 1 **Whiz-Bang Demonstrations,** The Dollar-Bill Bridge, Demo 37
		...in how models represent real objects or systems. ...examples of different ways models are used in science.	**Making Models,** Off to the Races! p. 627 **Datasheets for LabBook,** Off to the Races! Datasheet 2
		...in the importance of the International System of Units. ...rmine the appropriate units to use for particular ...easurements. ...ribe how area and density are derived quantities.	**Skill Builder,** Measuring Liquid Volume, p. 628 **Datasheets for LabBook,** Measuring Liquid Volume, Datasheet 3 **Skill Builder,** Coin Operated, p. 629 **Datasheets for LabBook,** Coin Operated, Datasheet 4 **Long-Term Projects & Research Ideas,** Project 51

TECHNOLOGY RESOURCES

Guided Reading Audio CD
English or Spanish, Chapter 1

Science Discovery Videodiscs
Image and Activity Bank with Lesson Plans: Science and the Constitution, Models and Predictions
Science Sleuths: The Traffic Accident

Scientists in Action, Remembering Richard Feynman, Segment 3
Multicultural Connection, Hopi Science, Segment 1

Classroom Management CD-ROM

Test Generator CD-ROM

3A Chapter 1 • The World of Physical Science

Annotated Teacher's Edition

Each Chapter Interleaf in the Annotated Teacher's Edition contains a comprehensive Chapter Organizer in chart form. The last column in this chart, Review and Assessment, clearly identifies the assessment components found in each section of a chapter. For example, the exact page numbers of Review questions and Self-Check questions found in the Pupil's Edition are shown. In addition, the page numbers of quiz questions, homework assignments, and alternative assessment options found only in the Annotated Teacher's Edition are listed.

The Chapter Interleaf also contains an End-of-Chapter Review and Assessment box that lists the location of important review and assessment items. Both of these handy references to assessment allow you to find the assessment materials you need without having to leaf through the entire chapter or an ancillary book.

Chapter 1 • The World of Physical Scien

CLASSROOM WORKSHEETS, TRANSPARENCIES, AND RESOURCES	SCIENCE INTEGRATION AND CONNECTIONS	REVIEW AND ASSESSMENT
Science Puzzlers, Twisters & Teasers, Worksheet 1 **Directed Reading Worksheet 1** **Science Skills Worksheet 8,** Reading a Science Textbook		
Directed Reading Worksheet 1, Section 1 **Science Skills Worksheet 30,** Hints for Oral Presentations	**Connect to Earth Science,** p. 7 in ATE **Multicultural Connection,** p. 7 in ATE **Cross-Disciplinary Focus,** p. 8 in ATE **Careers:** Electronics Engineer–Julie Williams-Byrd, p. 32	**Review,** p. 10 **Quiz,** p. 10 in ATE **Alternative Assessment,** p. 10 in ATE
Directed Reading Worksheet 1, Section 2 **Math Skills for Science Worksheet 16,** What Is a Ratio? **Transparency 201,** The Scientific Method **Reinforcement Worksheet 1,** The Plane Truth	**Math and More,** p. 13 in ATE **Cross-Disciplinary Focus,** p. 13 in ATE **Biology Connection,** p. 14 **Real-World Connection,** p. 17 in ATE **Multicultural Connection,** p. 18 in ATE **Holt Anthology of Science Fiction,** *Inspiration*	**Review,** p. 14 **Homework,** p. 15 in ATE **Self-Check,** p. 16 **Review,** p. 19 **Quiz,** p. 19 in ATE **Alternative Assessment,** p. 19 in ATE
Directed Reading Worksheet 1, Section 3	**Apply,** p. 21 **Real-World Connection,** p. 21 in ATE **Cross-Disciplinary Focus,** p. 21 in ATE **Connect to Earth Science,** p. 22 in ATE	**Homework,** pp. 20, 22 in ATE **Review,** p. 23 **Quiz,** p. 23 in ATE **Alternative Assessment,** p. 23 in ATE
Transparency 202, Common SI Units **Math Skills for Science Worksheet 27,** What Is SI? **Math Skills for Science Worksheet 30,** Finding Volume **Transparency 4,** Scale of Sizes **Directed Reading Worksheet 1,** Section 4 **Math Skills for Science Worksheet 29,** Finding Perimeter and Area **Critical Thinking Worksheet 1,** A Solar Solution **Science Skills Worksheet 9,** Safety Rules!	**Math and More,** p. 25 in ATE **Connect to Life Science,** p. 25 in ATE **MathBreak,** Using Area to Find Volume, p. 26	**Homework,** p. 26 in ATE **Review,** p. 27 **Quiz,** p. 27 in ATE **Alternative Assessment,** p. 27 in ATE

internetconnect

Holt, Rinehart and Winston On-line Resources

go.hrw.com

For worksheets and other teaching aids related to this chapter, visit the HRW Web site and type in the keyword: **HSTWPS**

National Science Teachers Association

www.scilinks.org

Encourage students to use the keywords listed on the Technology Highlights page to access information and resources on the **NSTA** Web site.

END-OF-CHAPTER REVIEW AND ASSESSMENT

Chapter Review in Study Guide
Vocabulary and Notes in Study Guide
Chapter Tests with Performance-Based Assessment, Chapter 1 Test
Chapter Tests with Performance-Based Assessment, Performance-Based Assessment 1
Concept Mapping Transparency 1

Chapter 1 • Chapter Organizer 3B

REVIEW AND ASSESSMENT

Review, p. 10
Quiz, p. 10 in ATE
Alternative Assessment, p. 10 in ATE

Review, p. 14
Homework, p. 15 in ATE
Self-Check, p. 16
Review, p. 19
Quiz, p. 19 in ATE
Alternative Assessment, p. 19 in ATE

Homework, pp. 20, 22 in ATE
Review, p. 23
Quiz, p. 23 in ATE
Alternative Assessment, p. 23 in ATE

Homework, p. 26 in ATE
Review, p. 27
Quiz, p. 27 in ATE
Alternative Assessment, p. 27 in ATE

END-OF-CHAPTER REVIEW AND ASSESSMENT

Chapter Review in Study Guide
Vocabulary and Notes in Study Guide
Chapter Tests with Performance-Based Assessment, Chapter 1 Test
Chapter Tests with Performance-Based Assessment, Performance-Based Assessment 1
Concept Mapping Transparency 1

Assessment Strategies

Ideally, assessment should be ongoing and should measure performance on exams and quality of classwork. Homework, lab work, and ScienceLog entries should all be factors in assigning grades.

Reliance on recall-based assessment strategies provides a limited view of students' abilities. Teachers who currently rely heavily on such assessment strategies may find it difficult at first to adopt new methods of assessment. However, once the transition is made, the reward—in the form of improved student performance and motivation—will more than offset the inconvenience.

ScienceLog Assessment

The Pupil's Edition provides many opportunities for students to demonstrate their understanding of specific concepts. At the beginning of each chapter, students are asked several questions and are encouraged to write in their ScienceLog about what they already know or think they know.

What Do You Think?

In your ScienceLog, try to answer the following questions based on what you already know:

1. What is physical science?
2. What are some steps scientists take to answer questions?
3. What purpose do models serve?

After students complete the chapter, they are given the opportunity to revise their entries. Here, students will confront any misconceptions they may have had in the beginning. In this way students actually assess their own prior knowledge and make adjustments accordingly. Although you should not grade these ScienceLog entries beyond checking that students have done them, viewing students' initial entries can give you a good idea of their understanding of the main concepts. Therefore, these entries can provide you with an excellent diagnostic tool to determine where to start your teaching and what concepts will need the most emphasis.

By checking your students' revised ScienceLog entries, you can get a good idea of the progress that has been made as well as what topics may need to be revisited to ensure understanding.

In addition, the ScienceLog serves as a companion notebook for most of the written work that is assigned throughout the program. You may want to have students hand in their entire ScienceLog periodically to check their work and progress.

Portfolio Assessment

Holt Science and Technology is ideally suited to the use of portfolios as one method of assessing your students' performance and accomplishments. For more information about portfolios and their use, please refer to the article "Assessment that Emphasizes Learning," on page 72. For help in assessing student portfolios, several specific checklists and rubrics are available in the *Assessment Checklists & Rubrics* booklet, which is also available on the *One-Stop Planner CD-ROM.*

Assessing Scientific, Psychomotor, and Communication Skills

Scientific knowledge and understanding are closely linked to the development of important process skills such as observing, measuring, graphing, writing, predicting, inferring, analyzing, and hypothesizing. The learning tasks in this program are designed to help develop these skills. As a teacher, you can assess such skill development by inspecting student work and by observing student performance.

The sample tables below are suitable models for evaluating student performance.

Assessing Environmental Awareness

This middle-school program was written with a commitment to environmental awareness. Many activities that promote such awareness are included, and the teacher is provided with suggestions on extending this theme through creative projects, cleanup or recycling projects, and so on. Tasks such as these promote environmental consciousness. The care that students take in carrying out these activities is a measure of their awareness of environmental issues.

Assessing Scientific Behavior

BEHAVIOR	POOR	SATISFACTORY	GOOD	VERY GOOD	EXCELLENT
Cooperates with others in small groups					
Observes and records observations					
etc.					

Assessing Technical Skills

TASK	YES	NO	UNCERTAIN
Is able to read thermometer correctly			
Is able to use spring scale to measure force			
etc.			

Assessing Scientific Attitudes

It can be useful to survey your students about the types of science-related hobbies and interests that they pursue outside of class. In a direct way, this provides feedback on the success of your school's science program. A successful science program is reflected in a student body with outside interests in science. Ask your students to keep a tally of any science-related activities they undertake outside of class. These could include reading or writing about science and technology, undertaking science-related projects, visiting museums, attending lectures on scientific and technological topics, and viewing science programs on television.

For developing and assessing individual students' interest in and attitudes toward science, the assignment of elective reading and independent projects is essential. Student work on elective projects should count as a significant part of overall assessment. This type of work provides the surest indication of a student's interest and proficiency in science, especially the student's ability to study, plan, and research independently.

A successful science program is reflected in a student body with outside interests in science.

A Balanced Assessment

The authors of this program recommend a balance between the different forms of assessment. As a general rule, the following proportions are suggested:

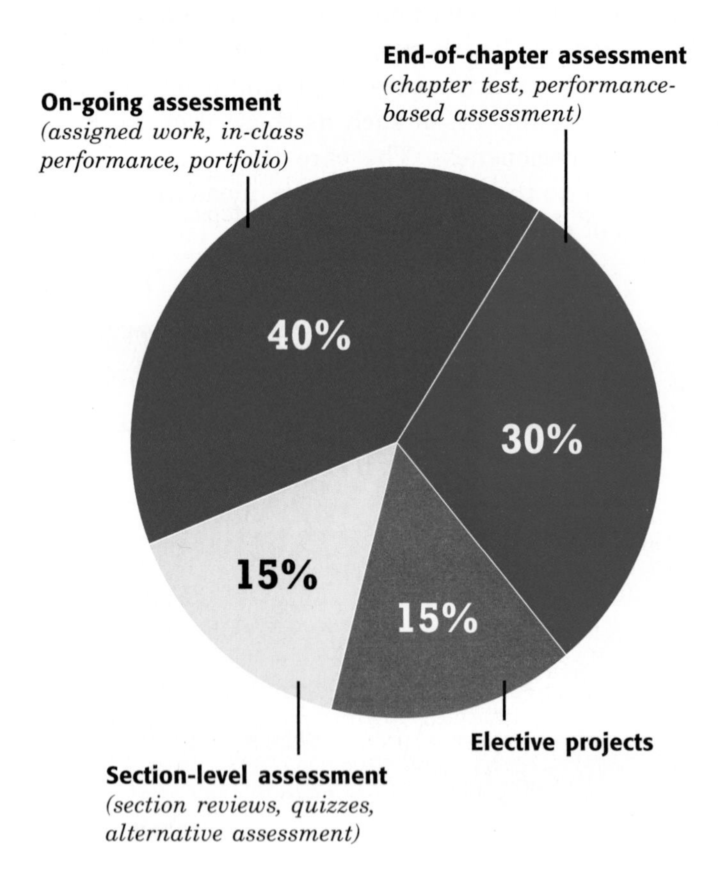

Assessing Science Projects

Undertaking a project provides students with a host of positive experiences. Students learn to organize, plan, and piece together many separate ideas and pieces of information into a coherent whole. Undertaking a project also allows students to experience the sense of accomplishment that comes from tackling and completing a difficult task. It has even been argued that no science education is complete without having undertaken and completed a major project.

Many students will resist the idea of undertaking a major project because they feel that it is too much work or that they are simply not up to the task. The following suggestions may help you overcome students' reluctance:

- Allow students to select their own project ideas.
- Encourage students to be creative.
- Provide a clear set of guidelines for developing and completing projects.
- Help students locate sources of information, including people in science-related fields who might advise students about their projects.
- Allow students the option of presenting their completed projects to the class.
- Emphasize the satisfaction students will derive from completing their projects.
- Inform students of the general areas in which assessment may be made, such as scientific thought, originality, and presentation.
- Do not emphasize the details of assessment. "Scoring points" should not be a major incentive.

Do not allow preconceived notions about how the project should be done to detract from the students' interest, enjoyment, and satisfaction in doing original work. Rather than forcing all students to fit their project work into a mold suited to scientific research, establish sets of criteria, as described in the sample rubrics on the following pages. Sharing your expectations with students at the outset of a project will help students rest assured that they understand the assignment and what it entails, and that their work will be fairly assessed.

Assessment Rubrics

Rubric for Reports and Presentations

Scientific Thought (40 Points Possible)				
40–36	**35–31**	**30–26**	**25–21**	**20–10**
Complete understanding of topic; topic extensively researched; variety of primary and secondary sources used and cited; proper and effective use of scientific vocabulary and terminology	Good understanding of topic; well researched; a variety of sources used and cited; good use of scientific vocabulary and terminology	Acceptable understanding of topic; adequate research evident; sources cited; adequate use of scientific terms	Poor understanding of topic; inadequate research; little use of scientific terms	Lacks an understanding of topic; very little research, if any; incorrect use of scientific terms
Oral Presentation (30 Points Possible)				
30–27	**26–23**	**22–19**	**18–16**	**15–5**
Clear; concise, engaging presentation; well supported by use of multisensory aids; scientific content effectively communicated to peer group	Well organized, interesting, confident presentation supported by multisensory aids; scientific content communicated to peer group	Presentation acceptable; only modestly effective in communicating science content to peer group	Presentation lacks clarity and organization; ineffective in communicating science content to peer group	Poor presentation; does not communicate science content to peer group
Exhibit or Display (30 Points Possible)				
30–27	**26–23**	**22–19**	**18–16**	**15–5**
Exhibit layout self–explanatory and successfully incorporates a multisensory approach; creative use of materials	Layout logical, concise, and can be followed easily; materials used in exhibit appropriately and effectively	Acceptable layout of exhibit; materials used appropriately	Organization of layout could be improved; better materials could have been chosen	Exhibit layout lacks organization and is difficult to understand; poor and ineffective use of materials

Rubric for Experiments

Scientific Thought (40 Points Possible)	
40–36	**35–5**
An attempt to design and conduct an experiment or project with all important variables controlled	An attempt to design an experiment or project but with inadequate control of significant variables

Originality (16 Points Possible)				
16–14	**13–11**	**10–8**	**7–5**	**4–2**
Original, resourceful, novel approach; creative design and use of equipment	Imaginative extension of standard approach and use of equipment	Standard approach and good treatment of current topic	Incomplete and unimaginative use of resources	Lacks creativity in both topic and resources

Presentation (24 Points Possible)				
24–21	**20–17**	**16–13**	**12–9**	**8–5**
Clear, concise, confident presentation; proper and effective use of vocabulary and terminology; complete understanding of topic; able to arrive at conclusions	Well–organized, clear presentation; good use of scientific vocabulary and terminology; good understanding of topic	Presentation acceptable; adequate use of scientific terms; acceptable understanding of topic	Presentation lacks clarity and organization; little use of scientific terms and vocabulary; poor understanding of topic	Poor presentation; cannot explain topic; scientific terminology lacking or confused; lacks understanding of topic

Exhibit (20 Points Possible)				
20–19	**18–16**	**15–13**	**12–11**	**10–6**
Exhibit layout self–explanatory and successfully incorporates a multisensory approach; creative and very effective use of materials	Layout logical, concise, and can be followed easily; materials used appropriately and effectively	Acceptable layout; materials used appropriately	Organization of layout could be improved; better materials could have been chosen	Layout lacks organization and is difficult to understand; poor and ineffective use of materials

Rubric for Technology Projects

Scientific Thought (40 Points Possible)				
40–36	**35–31**	**30–26**	**25–21**	**20–10**
An attempted design solution to a technical problem; the problem is significant and stated clearly; the solution reveals creative thought and imagination; underlying technical and scientific principles are very well understood	An attempted design solution to a technical problem; the solution may be a standard one for similar problems; underlying technical and scientific principles are recognized and understood	A working model; underlying technical and scientific principles are well understood; model is built from a standard blueprint or design	Model is built from a standard blueprint or design or from a kit; underlying technical and scientific principles are recognized but not necessarily understood	Model is built from a kit; underlying technical and scientific principles are not recognized or understood
Presentation (30 Points Possible)				
30–27	**26–23**	**22–19**	**18–16**	**15–5**
Clear, concise, confident presentation; proper and effective use of vocabulary and terminology; complete understanding of topic; able to extrapolate	Well-organized, clear presentation; good use of scientific vocabulary and terminology; good understanding of topic	Presentation acceptable; adequate use of scientific terms; acceptable understanding of topic	Presentation lacks clarity and organization; little use of scientific terms and vocabulary; poor understanding of topic	Poor presentation; cannot explain topic; scientific terminology lacking or confused; lacks understanding of topic
Exhibit (30 Points Possible)				
30–27	**26–23**	**22–19**	**18–16**	**15–5**
Exhibit layout self–explanatory and successfully incorporates a good sensory approach; creative and very effective use of material	Layout logical, concise and easy to follow; materials used in exhibit appropriately and effectively	Acceptable layout of exhibit; materials used appropriately	Organization of layout could be improved; better materials could have been chosen	Layout lacks organization and is difficult to understand; poor and ineffective use of materials

The Well-Managed Classroom[1]

by Harry K. Wong

Harry K. Wong *has over 35 years experience in classroom teaching. He lectures to educators and administrators around the world. His students have won more than 200 awards for academic excellence. With his wife, Rosemary, he authored* The First Days of School, *a best-selling book for teachers. Wong has received many awards, including the Outstanding Secondary School Teacher Award, the Outstanding Biology Teacher Award, and the Valley Forge Freedom's Foundation Teacher's Medal.*

Effective teachers typically have specific characteristics in common—positive expectations for student success, the ability to manage a classroom effectively, a knowledge of lesson design that leads to the students' mastery of lessons, and the drive to continuously learn about and grow within the teaching profession.

The First Days of School is a guidebook that my wife, Rosemary, and I wrote to help teachers everywhere be as efficient and effective as they can, regardless of their teaching style or level of experience. We have devoted a whole unit of the book to each important quality of an effective teacher. Of all these things, the principles of successful classroom management are probably the most important. Every student and teacher will be more successful in a well-managed classroom.

What Is Classroom Management?

Classroom management refers to all of the things that a teacher does to organize students, space, time, and materials so that learning can take place. This management includes fostering student involvement and cooperation in all classroom activities and establishing a productive working environment. Let's examine some characteristics of a well-managed classroom.

The Characteristics of a Well-Managed Classroom[2]

1. Students are deeply involved with their work, especially with academic, teacher-led instruction.
2. Students know what is expected of them and are generally successful.
3. There is relatively little wasted time, confusion, or disruption.
4. The climate of the classroom is work-oriented but relaxed and pleasant.

[1] Adapted from Wong, Harry K., and Rosemary T. Wong. *The First Days of School.* (Complete references can be found in *Section III: Continuing the Discussion,* page 139.)

[2] Sanford, et al., 1983.

Beginning the Year as an Effective Teacher

Establishing a well-managed classroom early in the school year (and in the teaching career if possible) can help a teacher avoid being part of the 40 percent each year who leave the profession discouraged and overwhelmed.

Contrary to what many people believe, the number-one problem in the classroom is not discipline—it is the lack of procedures and routines. Classroom management has nothing to do with discipline. You manage a store; you do not discipline a store. You manage a classroom; you do not discipline a classroom.

Behavior will rarely become a problem when effective teaching is already taking place. On the first day of school, you can begin teaching the procedures and routines that you will use. Never assume that you will have time to tackle bad behavior later. Becoming an effective teacher depends on classroom management—especially on the first day of school.

Classroom Management on the First Day of School

- Make sure your classroom is ready. Teachers who prepare their classrooms in advance maximize student learning and minimize student misbehavior. Readiness is the primary determinant of teacher effectiveness.
- Do everything possible to welcome the students and to make sure that they know where to go and how to get there on time.
- Keep in mind that what you do on the first day may determine how much respect and success you will have for the rest of the school year.
- Arrange student seating to maximize the accomplishment of the tasks and to minimize behavior problems. Assign students to their seats on the first day of school.

I love to stand at the door on the first day with a giant smile on my face, hand stuck out in an invitational pose, waiting for those "little darlings" to come down the hall.

Example of an Introduction

In order to have the respect and image that you desire in your class, you need to start off the year on the right foot. If you introduce yourself in a simple and positive manner, you will help the students feel comfortable in your class and communicate your positive expectations for the year. I love to stand at the door on the first day with a giant smile on my face, hand stuck out in an invitational pose, waiting for those "little darlings" to come down the hall. There are two major things you want to state at the outset on the first day of school: your name and your expectations. The following is an example:

Welcome. Welcome to another school year.

My name is Mr. Wong. There it is on the chalkboard. It is spelled W-O-N-G and is pronounced "wong." I would like to be addressed as Mr. Wong, please. Thank you.

I am looking forward to being your teacher this year. Relax. I have over 30 years' experience as a teacher. I am what is called an experienced, veteran teacher.

In addition, I go to workshops, conferences, in-service meetings, college classes, and seminars. I also read the professional journals and work together with my fellow teachers. I am a competent, knowledgeable, experienced, and professional teacher.

Also, I love to teach, and I am proud that I am a teacher. So relax. You are in good hands this year with me, Mr. Wong. You are going to have the greatest educational experience of your life. We will not only study (subject), but I will also share with you some life-skills traits that will help you to be successful in tomorrow's world.

I can assure you that if you should run into me at the shopping mall 25 years from now, you will say, "You were right, Mr. Wong. That was the most memorable, exciting, and fascinating class I ever had."

So, welcome!

Continuing the Year as an Effective Teacher

Procedures and routines facilitate classroom management. A procedure is not a discipline plan, nor is it a threat or an order. Rather, a procedure is a method or process for accomplishing things in the classroom—for example, what to do when entering the classroom, how to function in a lab group, or what to do when you have a question. A series of procedures and routines creates a structure for the classroom. When students know how the classroom operates, the class suffers fewer interruptions. A class with few interruptions is a class that advances learning.

Classroom Management All Year Long

- ◆ Organize a well-managed classroom in which students can learn in a task-oriented environment.
- ◆ Start the class by giving an assignment, not by taking roll. There is no need to involve the class in the roll-taking process.
- ◆ Post your assignments in the same place every day if you want your students to do them.
- ◆ Make sure that your grade record book shows the results and progress of each student at all times.
- ◆ Remember that a smooth-running class depends on your ability to teach procedures.
- ◆ Present your rules clearly, and provide reasonable explanations of the need for them. Write the rules down, and permanently post them in the classroom. Give them to students on paper or have the students copy them into their notebook.
- ◆ Keep in mind that rules are most effective when there are consequences to face if students break them and rewards if students follow them. When you see a violation of one of the rules, immediately and quietly give out the penalty as you continue with the lesson or classwork.

Classroom Procedures

As in real life, there are procedures in the classroom. Every time the teacher wants something done, there must be a procedure or a set of procedures. Some procedures that nearly every teacher must teach include the following:

- procedure for dismissal at the end of the period or day
- procedure for when students are absent
- procedure for quieting a class
- procedure for the beginning of the period or day
- procedure for students seeking help
- procedure for the movement of students and papers

Procedure for the Beginning of the Period or Day

An effective teacher always has the procedure or schedule posted or ready for distribution when the students arrive. Research has shown that effective classroom managers 1) have an assignment posted before the students enter and 2) have it posted in the same consistent location every day. This way, the students know that they are to get to work immediately upon entering the classroom.

Assignment

1. Put your backpack away.
2. Sharpen your pencil.
3. Organize your desk.
4. Complete the short assignment on the board.
5. Read quietly while waiting for the next instruction.

> *"I put an assignment on the board every single day before the students come into my classroom. I now have one of the smoothest running classrooms, and the students produce more for me now than at any other time. And I have been teaching for 14 years."*
>
> *Lisa McKuin, science teacher*
> *Alma High School*

The Student Who Is Absent

I have a procedure for roll taking and for students who are absent. I have three students trained to take the roll on a rotating basis. They do this while the students are completing their opening assignment.

If a student is absent, they complete a form that says, "Make-up work for Mr. Hockenberry," clip it to the work that has already been prepared, and place it in an envelope along one of the walls marked with the appropriate period.

A returning absent student does not come to see me. When absent students return, they obtain their work from the envelope. If they don't understand something, they ask one of the three students before coming to me for help. They seldom do, and class proceeds quickly with the lesson for the day.

Ed Hockenberry
Midlothian Middle School

Procedure for When Students Are Absent

Much class time can be wasted, and a class can become disrupted, if a teacher has to spend time gathering materials for a student who has been absent. Having a procedure for getting make-up work allows teachers to manage their classrooms effectively and puts the responsibility and accountability on the student.

Julie Guillory, a high school science teacher outside Houston, Texas, has a bulletin board with the daily worksheet and make-up work posted in envelopes. The students know where to find their work and what to do if they have been absent.

Teaching Classroom Procedures

Most behavior problems in the classroom are caused by the teacher's failure to teach students how to follow procedures. Teachers must learn how to effectively convey the procedures just as students must learn how to follow the procedures. Below is a summary of an effective method of teaching classroom procedures.

The Three-Step Approach to Teaching Classroom Procedures

1. **Explain:** State, explain, model, and demonstrate the procedure.
2. **Rehearse:** Rehearse and practice the procedure under your supervision.
3. **Reinforce:** Reteach, rehearse, practice, and reinforce the classroom procedure until it becomes a student habit or routine.

Effective teachers know what activities need to be done and have worked out the procedures for each of them. It is urgent that you have the procedures for each activity ready on the first day of school. Revise and hone these procedures year after year until they become models of efficiency.

Effective teachers spend a good deal of time during the first weeks of school introducing, teaching, modeling, and rehearsing procedures. Do not expect the students to learn all the procedures in one day. Behaviors must be taught, modeled, practiced, monitored, and retaught.

When procedures are performed correctly, there should be words of praise and smiles. Effective teachers reinforce the correct technique by having the student perform the technique over and over again, each time exhorting the student to do it better.

For example, if a student rushes into the classroom and pushes another student, ask him or her to return to the door and try again. Tell the student why. Give the student specific directions (walk quietly, don't push anyone, go directly to your seat, begin the work that is posted on the board, etc.), and be sure to use the student's name and say "please" and "thank you" to model respectful behavior.

Example of Teaching a Procedure: Quieting a Class

Do you yell, scream, and flick the lights to get your class quiet, with no good results? Quieting a class can be achieved with the following easy steps:

#1 Explain

Students, I have a procedure to get your undivided attention. You will see me stand here with my hand up. Or I may hit a bell because some of you will not be able to see my hand while you are working in a group. When you see my hand raised or hear a bell, the procedure is as follows:

1. Freeze.
2. Turn and face me, pay attention, and keep your eyes on me.

3. Be ready for instruction. I will have something to say.

Repeat, and look for class understanding.

Byron, please tell me the procedure when you see my hand raised or hear a bell.

Byron does so.

Yes, yes, yes, thank you, Byron.

Repeat this with several more students.

Is there anyone who does not understand or know what to do if you see my hand raised or hear a bell?

#2 Rehearse

Good, let's rehearse the procedure.

We will be working together this year, so let's get to know one another. Please look at the people to the right of you. You may have 2 minutes to introduce yourself and get acquainted.

At 2 minutes, hold up your hand and hit the bell, perhaps doing both this first time. Do not say a word. Carry out the procedure exactly as you plan to do it for the rest of the year. Be patient and wait until the class completes the three steps and is paying attention. Do not give up as you wait for the students to give you their undivided attention. Compliment them when you have their attention.

Thank you. You practiced the procedure correctly. Now let's try a different scenario. You will often find yourself out of your seat, working in groups or alone somewhere in the room.

Direct two students to stand at the pencil sharpener, two at the bookcase, and one at the computer. Then hold up your hand, and wait for the students to pay attention.

#3 Reinforce

Thank you. That was the correct procedure for what happens when I hold up my hand or ring the bell. Please do the same thing each time you see my hand raised or hear the bell.

Keep using the same wording if you want the students to practice the same routine.

Beginning Each and Every Day the Right Way

Just as it is easier to get control at the beginning of the year than it is to regain control if you've lost it, it is easier to start each class period with a quiet class than it is to quiet a noisy class. You are far less likely to ever have to worry about discipline problems if your class is continuously occupied. As you can see, a strong, positive start to the school year and the immediate implementation of a few simple procedures provide a structure that can help a teacher have a well-managed classroom. A well-managed classroom gives every student the chance to have one of the best learning experiences of his or her life. ■

The Top 10 Things New Teachers Should Know

by Catherine Wilcoxson, Ph.D.

Catherine Wilcoxson *is an associate professor of biology education in the department of biological sciences at Northern Arizona University, in Flagstaff, Arizona. Her doctorate, from the University of Nebraska at Lincoln, is in the area of curriculum and instruction. Prior to her position at Northern Arizona University, Wilcoxson was the project coordinator for the Mathematics/Science Framework Project for the Department of Education, in Lincoln, Nebraska. Wilcoxson also has 20 years' experience as a junior and senior high school science teacher. She has edited and reviewed chapters in math and science textbooks and lab manuals, as well as authored a number of articles in journals such as* American Biology Teacher *and* Science Scope.

As teachers begin their careers, they are usually excited about their first "real" teaching position. This feeling is followed by one of uncertainty. Do I have the knowledge and skills to be in total control of a classroom? What do I do when a student talks back to me? How will I decide what topics to teach? How should I assess students' progress? What is the best way to maintain discipline in my classroom? Will the students like me? These are all valid questions, and you will find that as you evolve as a teacher, you will return to them frequently. As a first-year teacher, you have the exciting challenge of teaching students and learning what it means to be an effective teacher. The tips in this article will give you a head start on this process by giving you some basic ideas to help you be a successful teacher and to help your students become better learners.

College courses are filled with theory and a number of instructional and learning strategies that you can use to become a successful teacher. They give you—the future teacher—many ideas that you may incorporate into your classroom. The big question is how to translate these ideas into practice. How do I decide which strategy to use when? What are some practical tips that will help me in my day-to-day practice? How can I successfully survive my first year of teaching?

Here are my top 10 hints for successful teaching that your college professors and texts probably did not mention.

1 **Don't worry so much about doing the right thing.** There is no one right way. You must determine what works best with your personality, your students, and school policies. It will help if you have read your school's student handbook. The guidelines in the handbook will help you avoid some of the more obvious pitfalls and give you some insight into the culture of your new school.

2 **Relax.** Get to know your students. Listen, empathize, and above all be patient. Your students are great people. Let them know that you care about them and their education. Also let the students know that you are not infallible—be comfortable enough with yourself that you can admit when you do not know an answer and that you are willing to learn with your students. Establish that "sense of presence" that conveys to students and anyone observing the classroom that you are confident, capable, organized, prepared, responsive to students' needs, and able to make the class engaging for all students. Remember, stop worrying about yourself and what you are doing. Think about your students and enjoy working with them.

Stop worrying about your abilities and enjoy your students.

3 **Teaching is hard work.** It is much more work than you ever thought possible. It is essential that you are organized and use your time efficiently. Beginning teachers are generally surprised at the amount of time it takes to plan effectively. Determining objectives, deciding on activities, collecting materials, grading papers, and making informed decisions take a tremendous amount of time. And just when you think you have everything planned and organized, there is an unexpected fire drill, student assembly, or early dismissal that forces you to adjust your plans. It is essential that beginning teachers give themselves sufficient time to plan as well as learn to be flexible.

Flexibility in time management is also essential to accommodate the diversity of students' background knowledge and the manner in which they learn to process information. Successful teachers are willing to stop and reteach if necessary or to modify lesson plans if it becomes clear that more or less time is required.

It will help if you thoroughly think through your day's activities and objectives, keep careful records and notes, and prepare to explain and give common everyday examples. Don't be afraid to ask for suggestions and opinions. Talk to colleagues. Most teachers are flattered that you asked.

4 **Standards cannot be ignored and should not be feared.** There are pressures of test scores and state standards that must be met. Standardized tests are often used to determine school and district performance, allocate funds, and gain public approval. These test scores may even be published. Look at the standards as an aid to good teaching. Standards help you decide what to teach, support your decisions about what and how to teach, convince the public that you are familiar with the field of science teaching, and focus curriculum so that you do not spend too much time on a favorite subject and ignore important concepts that must be taught. To relieve your anxiety about standards, offer to serve on the committee that implements standards in your school.

5 **Write objectives that focus on student learning, not on today's task.** For example, "Build a model of a dam" is a task while "Simulate the principle of how dams work" is what you really want students to learn. Focus on what you really expect students to learn and to be able to do. Thinking about the criteria for student success before giving an assignment will help you determine your expectations. Critics will say this leads to grade inflation, but I disagree. Almost all students want to succeed. Clearly stating your expectations upfront helps students to know what they need to do to be successful and helps to build student confidence in you as a fair teacher.

Expectations need to be realistic and challenging for all students. Creating a list of expectations that provides for a range of abilities and skill levels can be difficult; however, students of all abilities must have the opportunity to succeed. Having clear expectations (objectives) will help you focus your lesson plan by allowing you to ask yourself, "How does this lesson or activity bring my students closer to meeting expectations?"

6 Smile, and do so frequently. Forget the adage, "Don't smile until Christmas." Smiling and demonstrating a sense of humor will not compromise your authority. You may be the only adult who smiles or greets an individual student warmly today. You may never know the importance of those smiles or other expressions of caring. Remember, you do not know what has happened to your students before they entered your classroom. Maybe they have had a fight with their mom or dad or with a sister or brother before leaving for school; maybe they haven't eaten; maybe they are feeling sick or have been teased in the hallway. Give them a break. Make them happy to be in your classroom every day.

Smiling and demonstrating a sense of humor will not compromise your authority

Think about the difference in the following scenarios: A student has been absent from class for several days. As you greet students coming into the classroom, you might say in a stern and authoritarian voice "Tommy, you have been absent for three days. Your homework is in your folder. You have until Thursday to turn it in." Or as you greet the student, you smile and say, "Tommy, I'm so glad to see you. I was beginning to be concerned. You have been gone for three days. Your homework is in your folder. Please pick it up and complete it by Thursday. You may ask me or your fellow students for help." In both cases, you are the person in charge of the classroom. However, in one case, a positive classroom environment is established where a student feels important and cared for, while in the other, rules or procedure is more important than the child. Which classroom would you rather be in?

7 Respect your students. Treat your students as intelligent humans beings. Avoid categorizing them as lazy, bright, unorganized, etc. Rather, look for each student's unique qualities. Do not allow put-downs or sarcasm in your classroom. Show by example that every comment is valued and must be constructive or helpful to others. Students can be taught responsibility for their own actions by following your example. This means you should always be prepared, return papers promptly, and provide feedback that is quick, helpful, and thoughtful. Chances are that if you respect your students, they will respect you.

8 Believe in your students even if they don't believe in themselves. Treat students as valued members of society. Give them hope and confidence that they can develop as strong individuals. Giving students answers or taking over and doing an experiment or activity for them leads to learned helplessness. The students begin to feel incompetent and doubt their ability to succeed. It is also important that discipline is not perceived as being personal. The act is criticized, not the student who has erred in judgment. We are all learning acceptable modes of behavior. Help them learn.

9 Don't isolate yourself. New teachers are often overwhelmed with the demands of their first teaching job, fear of failure, and uncertainty about the best course of action. Compounding this situation can be a feeling of isolation and the need to talk about work and the problems you have encountered. It is essential that beginning teachers become proactive in developing a support system. Often experienced teachers become entrenched in a day-to-day routine and are so busy with their own concerns that they do not realize that a beginning teacher might need help or support. Take the initiative to develop personal and professional relationships in your school. If possible, establish a mentor relationship with an experienced teacher. A mentor relationship encourages you to cooperatively seek solutions, increases your awareness of alternatives, provides a sounding board to vent frustrations, and allows you to learn from the experiences of a colleague.

Don't be afraid to learn from students. Students have such great ideas and creative solutions to problems!

10 It is not appropriate to teach the way we were taught. Chances are that your students are growing up in a world completely different from the one you grew up in. Technology, lifestyles, economics, and many other factors have created a different world than when you were in school. Students at this level are still curious, imaginative, enthusiastic, and talkative. The classroom environment tends to be very lively. However, as a first-year teacher, you may be surprised to learn that students are not as you remember yourself as a middle school student. Today's students seem less motivated to learn and more interested in other things such as listening to music, going to the mall, or playing video games. You can capitalize on your students' energy and enthusiasm by making connections between what you teach in the classroom and your students' everyday lives. Before you can do this, you need to know what students' lives are like outside the classroom.

A good way to learn about your students is to allow them to choose their own topics for projects. This gives you the opportunity to see what your students are interested in and helps make the learning relevant. An added benefit is that when students are involved in deciding what to study and how to demonstrate their understanding, they develop a sense of ownership and generally try to exceed your expectations.

In conclusion, effective teachers are constantly learning from their successes and failures. To be effective, teachers must look back on their practices and assess what works well, what doesn't, and how they might improve as teachers. Keeping a journal is an excellent means of keeping track of your performance, and it can provide valuable information for your own professional development.

Best wishes for many years of successful teaching!

Catherine

Yes, Teaching Students to ARGUE Is a Good Idea… No, I'm Not CRAZY!

by Valerie Goff Whitecap, Ph.D.

Valerie Goff Whitecap *received her Ph.D. in rhetoric and communication from the University of Pittsburgh. For the past 27 years, she has taught communication arts at Fort Couch Middle School, in Upper St. Clair, Pennsylvania. Whitecap has also taught and lectured extensively in the areas of public speaking, nonverbal communication, rhetoric, acting, intercultural communication, argumentation, story and script writing, and interpersonal communication. Since 1976, she has been an adjunct professor at Carlow College, in Pittsburgh, Pennsylvania; an assistant professor at Westminster College, in New Wilmington, Pennsylvania; and an instructor at the University of Pittsburgh, in Pittsburgh, Pennsylvania.*

One Halloween, Laura entered my classroom with her hair sprayed green.[1] Looking up, Charlie said, "You know, Laura, I don't think it's appropriate to come to school with your hair sprayed green." My ears perked up. I thought this might be the start of a rousing discussion on appropriate dress, peer pressure, or maybe even what constituted classroom distraction. Instead, Laura began yelling and swearing at Charlie, who appeared shocked. He didn't know how to respond to her. Before I knew it, the class had divided into two factions, each trying to outyell the other. Soon I restored calm and began the day's lesson, but I kept thinking about this random confrontation. I was struck that Laura didn't know how to deal with someone who disagreed with her and that Charlie didn't have the skills to deflect her anger or clearly state his position.

[1] Students' names have been changed.

An Interesting Theory

Shortly afterward, I heard Andrew Rancer discuss a theory on constructive argument that was originally developed by Dominic Infante (1988). Rancer had conducted numerous studies with adults using this theory and had achieved considerable success.

After receiving training on effective arguing, people were viewed as having higher credibility, they got higher job ratings and better raises, and they lowered their use of verbal aggression.

After receiving training on effective arguing, people were viewed as having higher credibility, they got higher job ratings and better raises, and they lowered their use of verbal aggression. Studies showed that the quality of decisions and creativity in decisions were also enhanced in groups that applied the principles of constructive argument.[2]

I had learned that all research on this topic was conducted with adults, but I knew that middle school students could benefit from this type of training. So I decided to teach my own students how to use constructive arguing.

Taking It to the Classroom

Redefining Familiar Terms

My first step was to adapt the materials used with adults to make them appropriate for middle school students. Then the training began. (For a full discussion of the adaptation, teaching methods, testing, and results of this research, see Rancer, Whitecap, et al., 1997.) Rancer and I decided on several points to emphasize to students during the training. The first involved the word *argument*. Almost everyone, including students, has a negative connotation of the word. The reason is that in English we often use the word interchangeably with words such as *fight*, *battle*, or *talking back*. In contrast to these popular conceptions of the word, we defined an *argument* as an opportunity to state and define a position in order to change the position of another.

Arguments Attack Ideas

The main point we wanted to stress to students was the following: Arguments state and attack ideas, not people. I told my students to think of this analogy: If I have a quiver of arrows and I am fighting with you, I try to hit you with the arrows. My purpose is to hurt you. If we are arguing, we both put out ideas between us, and we aim our arrows at the ideas, not at each other. Arguments are never intended to hurt anyone personally.

[2] Rancer, Whitecap, et al., 1997. (Complete references can be found in *Section III: Continuing the Discussion,* page 139.)

Arguing Is Not Fighting

The second important point that we targeted involved distinguishing between arguing and fighting. Often, children learn how to fight and win (or get the last word) by fighting with their siblings or friends. Then they use the same "skills" when trying to express their opinions with other peers and adults. From the perspective of an adult, these "skills" come across as "talking back," being contentious, or whining. It is important to emphasize to students that the skills they need to argue effectively are different from those necessary to "win" a fight with a sibling.

For an argument to be effective, all participants must understand the difference between a fight and an argument. They must agree to use language as a tool to further the argument, not as a weapon. For instance, in the confrontation between Laura and Charlie, we see from Laura's anger that she took Charlie's words as a personal attack. However, the arrow Charlie shot at the appropriateness of her hair color wasn't aimed at her. He hadn't said, "Laura, you are stupid for dyeing your hair green." She, however, responded as if he had aimed at her and fired back with language meant to hurt. Her words, or "locus of attack," did not fit within the bounds of constructive arguing, which only allows the attack of ideas. The ability of a student to find an appropriate "locus of attack" is essential to the student's success in constructive arguing.

They must agree to use language as a tool to further the argument, not as a weapon.

Addressing Verbal Aggression

The second prong of the training Rancer and I developed for students focused on gaining an understanding of verbal aggression. Verbal aggression involves the use of words to attack someone else or to elicit a strong reaction. Verbal aggression usually takes the form of teasing, insulting, swearing, or name calling. It is such an integral part of the way middle schoolers interact that they often don't see the behavior as destructive. In the case of Laura and Charlie, their disagreement became very loud and could have led to physical aggression if I had not intervened. To argue effectively, students must be able to distinguish verbally aggressive statements from statements that attack ideas only, and they must be willing to avoid using verbal aggression in their arguments.

Essential Strategies

Constructive argument requires a new way of thinking about disagreements. With a clear understanding of the difference between fighting and arguing as well as the ability to avoid verbal aggression, students are ready to have a constructive argument. To help themselves keep the correct locus of attack when arguing, students can follow a "thinking outline" that clarifies the questions and steps. The main steps are shown at right.

Steps of Constructive Arguing

1. Define the Problem
The first step to a constructive argument is defining what point or points will be argued. In this way, all participants can begin with the same locus of attack. For example, Laura and Charlie could have defined the central point of their disagreement as, "Student dress could disrupt the class by distracting or offending classmates and teachers."

2. Identify the Cause
It is also important to know whether the problem is caused by something the opponents can solve between themselves or whether the problem is really a part of a larger issue. Charlie needs to make clear what his real objection to Laura's hair is, and Laura needs to make clear what her problem is with Charlie's objection. Is Laura concerned about losing her freedom of expression? Is Charlie concerned with classroom distractions? An unclear school policy on dress code may be the cause of the problem.

3. Offer Alternative Solutions
The next step is to come up with several alternative solutions to the problem. The participants can evaluate those solutions and decide which solution is the best. For example, Laura and Charlie might have discussed several solutions, such as changing the school dress code, ignoring Laura's hair, or appointing a committee of students to recommend a new dress code.

4. Consider the Consequences
Next the opponents think about the consequences of the chosen solution. What good or bad things might happen if the proposed solution is adopted? For instance, a positive outcome of changing the school dress code to prohibit green hair might be that students would not be subjected to distracting appearances. A negative outcome might be that students would not be able to freely express themselves.

What makes this form of arguing so much different from the type of arguing that students are accustomed to? The main difference is that constructive arguing has a clear set of rules.

Rules for Constructive Arguing

LISTEN. Instead of thinking about what you want to say back, listen to what your opponent is saying and don't interrupt.

SAY EVERYTHING YOU WANT TO SAY, THEN STOP. When it is your turn, your opponent cannot interrupt you either, so say everything you want to say about the topic. Then stop talking.

DON'T REPEAT YOURSELF. When you run out of things to say, don't repeat what you have already said. "Is too, is not" bantering does not make a point any stronger.

NO DUMPING—STATE THE PROBLEM AND STICK TO IT. Don't dump all of your stored-up grievances at once. In other words, don't try to solve all of the problems of the world at once or drag in every example you can think of. Argue each point all the way through, one at a time.

ATTACK IDEAS, NOT PEOPLE. Arguments shouldn't hurt. Rather than saying to an opponent, "Boy, you have a stupid idea," try, "You have some interesting ideas. Let's see how we can make them better."

IF YOU CANNOT RESOLVE THE PROBLEM NOW, SET A TIME TO MEET AGAIN. Postponing the discussion can give you time to think further about what your opponent said and can give both of you a chance to think of alternative solutions.

IF YOUR OPPONENT USES VERBAL AGGRESSION, WALK AWAY. Walking away is often the most difficult response for students and adults alike. It is important to remember that the goal is to win an argument, not a fight. If a participant uses verbal aggression, say, "I'm willing to discuss this problem, let me know when you're ready." Then walk away.

Progress Report

I have been training seventh graders in constructive arguing for 3 years and have no plans to stop. My classes include gifted as well as challenged students, and all of my students have had some success. With their training, my students are able to think of more things to say, identify and avoid aggressive statements, and use logic in their arguments. I recently asked 58 students who had just completed the training, "What did you think of the argument program?" A total of 88 percent of the students thought the program was good and would help them. The smaller percentage said that they would like more training to better understand the method, had trouble exercising the patience or taking the time to think about using the method, or got too mad to think calmly and apply the method.

Give Arguing a Chance

Students today are more likely to demand to be heard instead of just willing to be seen. Rather than decrying their outspoken and sometimes disrespectful behavior, why not train them to use their voices to effectively present their ideas? Often middle school students simply need to be taught how to present their ideas so that others will listen to them.

Teaching students to argue constructively will not end fights in the classroom or outside of it. It will, however, give students valuable experience in exercising self-control and applying logical-thinking skills.

A question that teachers often ask me is, "Can students remember what they were taught and use it when it is really needed?" As with any worthwhile skill, arguing must be practiced before it becomes an automatic behavior. Review is important, but often a single word or phrase from me is enough to get my students to remember what to say or what not to say. I was pleased recently to overhear two eighth-grade boys in the hall. Ted said to Jon, "That's verbal aggression, and I don't have to listen to it now. Let's talk later when you're cool." Then Ted walked away.

Teaching students to argue constructively will not end fights in the classroom or outside of it. It will, however, give students valuable experience in exercising self-control and applying logical-thinking skills. The more students practice constructive arguing, the less likely they will be to fight and the more likely they will listen to and understand alternative viewpoints. Overall, I have found my students' discussions and disagreements to be more fruitful and peaceful. And after over 20 years in a middle school classroom, I can say that a little peace is a very welcome thing! ■

Understanding AGGRESSIVE Communication

by Andrew S. Rancer, Ph.D.

Andrew S. Rancer *is a professor of communication studies at the University of Akron, in Akron, Ohio. He received his Ph.D. in communication from Kent State University in 1979 and has since taught courses in communication theory, nonverbal communication, persuasion and attitude change, among other topics. The co-author of* Building Communication Theory *and the author of numerous professional journal articles, he has been researching and teaching others about interpersonal communication for more than 20 years.*

If you review your interactions with the people you encounter from day to day, you will probably recall instances in which your communication was marked by disagreement. In these interactions you and your adversary may have viewed the world in quite different ways. In some of these conflicts, you may have been engaged in an "argument" that you found stimulating, exciting, exhilarating, and possibly even fun. You may recall the excitement you felt upon successfully convincing someone that your position on an issue was the stronger one. These feelings of excitement, interest, and enjoyment may have led you to believe that arguing is a very positive activity.

On the other hand, you may recall instances in which these arguments were anything but enjoyable. That is, the argument led to feelings of anger, hurt, confusion, or embarrassment and may have even led to the termination of the relationship. Perhaps you can recall an example of an argument that became so destructive that it quickly turned into name-calling and may have even ended with the participants exhibiting physical aggression. These situations may have led you to conclude that arguing is a destructive and unsatisfying form of communication that should be avoided at all costs, even if it means suppressing your true feelings and yielding to another person's wishes. Those of us who research and study human communication behavior are perplexed by the seeming unfavorable attitudes that many people hold regarding the term arguing, because our discipline has advanced, since antiquity, that arguing is a constructive, important, and necessary form of communication.

As part of a conflict-management curriculum, we advocate the inclusion of a component on arguing constructively, which includes improving individuals' attitudes about and skill in argumentative communication.[1] This component makes use of an extensive body of research in the area of "aggressive communication predispositions." Research

[1] Rancer et al., 1997. (Complete references can be found in *Section III: Continuing the Discussion,* page 140.)

and theory-building in the area of aggressive communication was initiated and developed by Professor Dominic A. Infante, Professor Emeritus at Kent State University; myself; and Professor Charles J. Wigley III, of Canisius College.

Aggressive Communication and Predispositions

Those of us who research and study human communication behavior have come to understand aggressive communication as distinguished by a few specific behaviors. Aggressive communication involves one person applying force to another, typically with a high level of arousal. Participants engaged in aggressive communication are usually more active than passive, and they often adopt "attack" and "defend" modes of thinking and action. These types of behavior are essential for successfully resolving a conflict, though they can be used destructively as well as constructively.

Aggressive communication is most often controlled by four predispositions that interact with environmental factors to influence an individual's approach to conflict resolution. These four predispositions are classified as either constructive or destructive.

AGGRESSIVE COMMUNICATION PREDISPOSITIONS

Constructive	Destructive
Assertiveness	Hostility
Argumentativeness	Verbal aggressiveness

Constructive Predispositions Assertiveness and argumentativeness are viewed as constructive predispositions. Assertiveness includes characteristics of personal dominance, firmness, forcefulness, and the use of assertive behavior to achieve personal goals. Argumentativeness involves the use of reasoning to defend personal positions on controversial issues while attacking the positions of adversaries.[2] Argumentativeness can be understood as a subset of assertiveness; all argument is assertive, but not all assertiveness involves argument (e.g., a request). The communication discipline advocates the development of these two constructive traits in individuals. Time after time, research has shown that individuals who approach conflict from an argumentative stance are seen as more credible, eloquent, creative, and self-assured and are more likely to be viewed as leaders.

Destructive Predispositions Hostility and verbal aggressiveness are viewed as destructive predispositions. Hostility is characterized by the expression of negativity, resentment, and suspicion. Verbal aggressiveness is an assault on the self-concept, rather than the position, of an adversary. Individuals

[2] Infante and Rancer, 1982.

typically engage in verbal aggression in order to inflict psychological pain, such as humiliation, embarrassment, and other negative feelings about the self.[3] Compared with argumentative individuals, those who are verbally aggressive are seen as less credible, tend to suffer more strained relations with others, and resort to physical aggression and interpersonal violence more often.

Causes of Verbal Aggressiveness

Research has suggested a number of factors that may lead to a predisposition for verbal aggressiveness. One of the factors is repressed hostility. Individuals who were emotionally scarred by verbal aggression and hostility at a young age tend to demonstrate similar behaviors later in life. Because they were too young or lacked the power to reciprocate, they suppressed the hostility and have come to verbally aggress against those who remind them of the original source of hurt.

Social learning is also responsible for much verbal aggression. We learn to be verbally aggressive from various environmental forces, including our culture, our social group, our family, our friends, and the mass media. People reared in an environment of verbal aggression are more likely to exhibit this type of communication behavior.

Disdain is another common cause of verbal aggression. If we severely dislike someone, we are more likely to verbally aggress against him or her. While we generally try to ignore those we disdain, unavoidable (or even intentional) confrontations with them can rouse the ugliest verbal aggression in us.

Finally, many people resort to verbal aggression in order to compensate for a deficiency in argumentative skills. During conflict episodes, these individuals quickly use up their weak arguments only to find that their position is still not accepted. Because they find themselves in the "attack" and "defend" modes, they feel forced to use verbal aggression as a last resort.

We learn to be verbally aggressive from various environmental forces, including our culture, our social group, our family, our friends, and the mass media.

Types of Verbal Aggression

A taxonomy of verbally aggressive messages includes character and competence attacks, disconfirmation, physical appearance attacks, racial epithets, teasing, ridicule, threats, cursing, negative comparisons, and nonverbal aggression (e.g., rolling the eyes, gritting the teeth, looks of disdain, "flipping the bird"). All of these types of aggression are considered attacks on an adversary's self-concept and contribute little to nothing to the resolution of conflict. Often, in fact, they escalate the conflict, sometimes to the point of physical violence.

Reducing Verbal Aggression and Enhancing Argumentativeness

If we wish to resolve conflicts harmoniously and constructively, we must strive for a more argumentative and less verbally aggressive approach to conflict management. Extensive research on aggressive communication techniques and behaviors has led communication researchers to a set of recommendations that we believe can facilitate conflict resolution. These recommendations emphasize preventing verbal aggression from occurring, developing ways to reduce the effects of verbal aggression when it does occur, and providing some skills that can be used in effective argumentation. D.A. Infante, in "Teaching Students to Understand and Control Verbal Aggression,"[4] offers seven recommendations, shown in the chart on page 53.

[3] Infante and Wigley, 1986.

[4] Infante, 1995.

Conflict Management Techniques

Avoid Instruct students to avoid interacting with individuals who are known to be verbally aggressive. We should teach students how to detect behaviors (both verbal and nonverbal) that indicate verbal aggressiveness.

Be Polite If students are placed in situations in which conflict is apparent, they should allow their adversary to speak without interruption, try to use a calm delivery, use empathy (the ability to feel what another is feeling), allow the opponent to "save face" (stop short of humiliating the individual), and reaffirm the adversary's sense of competence.

Define the Argument Make sure that adversaries understand exactly what they are arguing about. One person may be arguing about one thing while the other is arguing about something else. In order for the argument to be resolved, both parties must agree on the proposition they are arguing.

Recognize Shifts from Constructive to Destructive Help students detect when a proposition has changed from an argumentative one into a verbally aggressive one (e.g., "We were arguing about who should wash the dishes, but you changed it to how selfish I am. Which one do you want to argue?").

Argue, Don't Aggress Teach students that verbally aggressive messages can be turned around and treated as "arguments." This allows students to use several strategies that are essentially argumentative and therefore constructive in nature. A verbally aggressive message is usually based on emotion, illogical claims and warrants, and insupportable data. It is usually easy to refute verbal aggression (e.g., "You say I don't know anything about football. That's not true, and this is why."). The student can attack the reasoning or evidence of the supposed argument and refute the adversary's claim.[5]

Be Aware Make students aware of situations and behaviors that are likely to stimulate verbal aggression, such as personal rejection, "hitting below the belt," non-negotiation, and "gunny-sacking" (storing up grievances and unloading them all at once). Once a student is aware of these situations and behaviors, he or she can try to avoid them.

Know When to Stop Teach students that if the verbal aggression continues, they can simply stop communication (e.g., "If you don't stop calling me stupid, I'm going home.").

[5] Infante, 1988.

Conclusion

Our research suggests that verbal aggression is a destructive form of communication, while arguing is constructive. We believe that improving individuals' argumentative skills can be employed as a strategy to prevent verbal aggression and to assist in productive conflict management. In order to accomplish this, students must learn the differences between these two forms of aggressive communication, and they must also learn key behaviors associated with arguing constructively. This brief review has attempted to introduce both.

Strategies for Improving Student Behavior

by M. Lee Manning, Ph.D.

M. Lee Manning *is a professor in the Department of Educational Curriculum and Instruction at Old Dominion University, in Norfolk, Virginia. His research and writing experience spans more than 20 years. In that time he has acquired national recognition in the areas of middle school education and multicultural education. Manning has authored over 150 journal articles and 10 books, some of which have been adopted by American universities as textbooks. He is also a regular presenter at national conferences on education. His books in progress include* Teaching in the Middle School, Multicultural Education of Children and Adolescents, *and* Teaching Learners At-Risk.

During the last 10–12 years, research on effective teaching has provided considerable insight into how teachers' attitudes and strategies influence academic achievement.[1] Many practicing educators have discovered key teaching practices that contribute positively to teaching effectiveness, time on task, and academic achievement. Often, though, teachers do not realize the potential these same techniques hold for improving classroom *behavior* as well. This paper reviews several studies of effective teaching practices to determine their implications for classroom management.

Selected Studies of Effective Teaching

The findings of researchers such as Brophy, Walberg, Porter, and Good have major implications for middle school educators seeking to improve student behavior.[2] The techniques for more effective teaching that the researchers have explored are outlined below. The studies key in on the following teaching behaviors:

- ✔ **"With-it-ness"**
- ✔ **Productive time on task**
- ✔ **Informing, teaching, and monitoring**
- ✔ **Putting learning first**

[1] Brophy, 1983; Brophy and Good, 1986; Walberg, 1988. (Complete references can be found in *Section III: Continuing the Discussion,* page 140.)

[2] Brophy, 1983; Brophy and Good, 1986; Walberg, 1988.

"With-it-ness"

Effective teachers demonstrate "with-it-ness." This means they work at becoming as aware as possible of all events and student behaviors in the classroom and closely monitor classroom activities.

What do teachers do to actually demonstrate with-it-ness? Brophy found that specific with-it behaviors are confronting problems before they become disruptive, monitoring all classroom activities, and stationing where students can be seen at all times. As a result, students perceive that their teacher is aware of their behavior and that he or she can detect inappropriate behaviors early and accurately.

In accordance with the old adage that teachers have "eyes in the back of their head," teachers who demonstrate with-it-ness are as aware as possible. While helping an individual student with seat work, a with-it teacher monitors the rest of the class, acknowledges other requests for assistance, handles disruptions, and keeps track of time. During a discussion, he or she listens to students' answers, watches other students for signs of comprehension or confusion, formulates the next question, and scans the class for possible misbehaviors. At the same time, the teacher attends to the pace of the discussion, the sequence of selecting students to answer questions, the relevance and quality of answers, and the logical development of content. When students are divided into small groups and the number of simultaneous events increases, the with-it teacher monitors and regulates several different activities at once.[3]

Mr. P. For example, Mr. P., a middle school English teacher, demonstrates several with-it behaviors, all of which contribute to his effectiveness as a classroom manager. He exhibits confidence in his ability to manage and teach his class of 28 seventh graders. He rarely experiences behavior problems because his students realize he is aware and in control. For instance, while simultaneously scanning the classroom and assisting a small group, Mr. P saw Sean and Darryl begin to scuffle. He quickly and firmly reminded them to get back on task, yet he didn't leave the small group. As a result of his with-it teaching behaviors, Mr. P has had very few behavior problems in his classes.

He rarely experiences behavior problems because his students realize he is aware and in control.

[3] Doyle, 1986.

Productive Time on Task

Walberg proposed "productive time" to be time spent on lessons adapted to learners' needs and interests, rather than just engaged time, which involves tasks designed to keep students busy and quiet.[4] Also, total immersion (cramming and intensive courses) can produce impressive results, but direct instruction interspersed with other teaching-learning activities proves more time efficient. Walberg summarized that only a fraction of engaged time proves productive since conventional "whole-group" instruction cannot accommodate the vast differences in individual learning rates and prior knowledge.

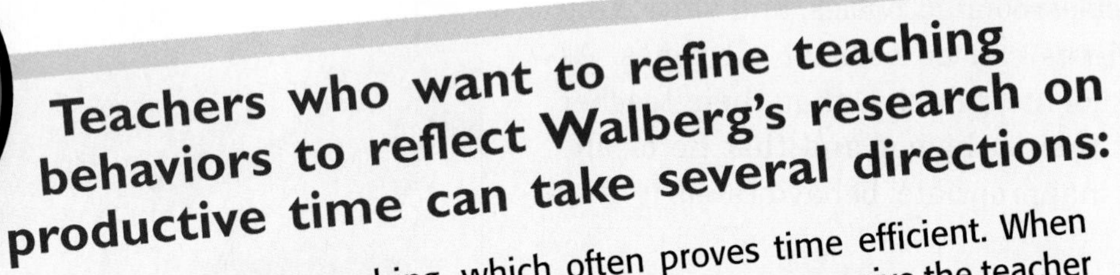

Teachers who want to refine teaching behaviors to reflect Walberg's research on productive time can take several directions:

- Use one-on-one teaching, which often proves time efficient. When teachers work directly with students, the students perceive the teacher as a methodical leader who carefully plans every lesson.
- Use well-paced learning activities combined with other school activities. When teachers monitor the pace and variety of lessons, students' frustration levels decrease, as does their perceived need to "cram."
- Use instructional activities that address a specific learning objective that learners consider appropriate and worthwhile. Students can detect when work is designed solely to maintain order and quietness.
- Use learning goals and instructional activities that accurately reflect young adolescents' prior achievement, age, or stage of maturation. Goals and activities that reflect students' accomplishments and self-esteem can lessen their need to misbehave for recognition.
- Use the school day productively by making a professional commitment to these goals.

Mrs. L. Mrs. L., a middle school science teacher, believes that all work should be as productive as possible. Therefore, students never do the same class work or homework twice unless a student's achievement suggests an actual need for it. Similarly, she models aptitudes and behaviors that reflect effective use of time and efforts. Mrs. L. uses large group instruction because it allows her to convey information to a large group of learners. Then she adapts instruction to learners' individual grade levels. Believing assignments are productive work instead of busy work, her students rarely misbehave and usually demonstrate productive time on task.

---

[4] Walberg, 1988.

Informing, Teaching, and Monitoring

Effective teachers clearly inform students of intended objectives, teach them expected learning strategies, and monitor learning progress. Research by Porter and Brophy provides a conceptual framework for effective teaching.[5] The chart at right highlights some of Porter and Brophy's key points. Teachers should also take time to reflect about instructional and management practices to determine how these practices might actually contribute to student misbehavior.

Brophy and Porter's research has a number of implications for middle school educators working to improve the behavior of 10–14 year olds. Their findings prove especially helpful because their synthesis focuses on specific teaching behaviors that can be translated into practice.

Ms. J. Ms. J., a third-year middle school teacher, always prepares her students for learning and thus improves the chances of positive student behaviors. She states one or two clear objectives at the beginning of the class, explains to students what they will learn and why, and makes sure they know appropriate learning strategies, i.e., cooperative learning or inquiry method. Only after she "primes" students for learning does she allow them to begin work. Ms. J. experiences few behavior problems because her students understand what to learn, recognize the importance of learning activities, and employ appropriate strategies to meet the learning objectives. Her students usually do not misbehave due to frustration because she constantly monitors the class for signs of confusion.

[5] Porter and Brophy, 1988.

Teaching behaviors that facilitate classroom management include the following:

- being clear about instructional goals, activities, and expectations to students, as well as the reasons behind them
- making expert use of curricular materials and instructional methods in order to reduce student confusion about goals and methods
- teaching students metacognitive strategies and providing opportunities to use them
- monitoring students' understanding by offering regular and appropriate feedback to help students realize that the teacher knows their actions and progress; this reduces the likelihood that students will engage in inappropriate behaviors
- getting to know students' strengths, weaknesses, and learning needs in order to better address the needs of each student

Steps to Promote Classroom Management

1. Plan rules and procedures in advance.
2. Establish clear rules and procedures when needed.
3. Allow students to assume responsibility for behavior.
4. Encourage teacher/student cooperation.
5. Minimize disruptions and delays.
6. Plan independent lessons and group lessons.

Putting Learning First

Good and Brophy reported several effective teacher behaviors that can help with classroom management.[6] Maximum achievement occurs when teachers do the following:

- *emphasize academic instruction*
- *expect all students to master the curriculum*
- *allocate the most time to curriculum-related activities*
- *assign seat work that constitutes a meaningful activity related specifically to the objectives and is at the appropriate level of difficulty*

Good and Brophy[7] suggested that teachers take specific and planned steps to promote classroom management. They based these steps on several assumptions, all of which have direct relevance to effective classroom management. First, students will more likely follow rules they understand. Second, students engaged in meaningful activities geared to their interests will be less likely to engage in disruptive behaviors. Third, students should see the learning environment as positive and conducive to behaving appropriately. Fourth, students need to develop self-control rather than have the teacher exert control over them.

Educators who recognize that their teaching behaviors and attitudes have a powerful effect on how students behave and think will find that Good and Brophy's conclusions provide a wealth of information and insights for effective classroom management. Undoubtedly, a number of causes (e.g., personal and familial problems as well as peer pressure) contribute to student misbehavior. However, perceptive educators also realize their powerful influences on students and on the teaching-learning environment.

---

[6] Brophy and Good, 1986.

[7] Good and Brophy, 1994.

A Sumary of Specific Behaviors What specific teaching behaviors have the potential for promoting classroom management? Good and Brophy's conclusions can be summarized as follows:

- *Teachers with high expectations for students' academic achievement and behavior often reduce the likelihood of students engaging in disruptive behaviors. This is especially true when students perceive that learning activities are geared to their grade and interest level.*
- *When teachers minimize delays in teaching-learning activities and provide work that reduces frustration, students will have less time to talk, walk around the classroom, and otherwise use time unproductively.*
- *When teachers ask questions at the appropriate difficulty level, wait a sufficient amount of time, and respond appropriately to incorrect/correct answers, they demonstrate awareness of student behaviors. As a result, students' attention remains on task.*

Mr. B. Mr. B. teaches sixth graders in an inner-city middle school known for low academic achievement and a high prevalence of behavior problems. Compared with other teachers in the school, however, Mr. B. experiences few behavior problems. From the first day of school, Mr. B. makes his expectations clear regarding academic achievement and appropriate behavior. He carefully does his part to promote appropriate behavior—he uses the students' instructional time effectively, insists upon academic achievement, teaches respect for self and others (he explains classroom rules and accompanying reasons), and makes sure teaching-learning experiences reflect students' ability levels. For instance, Bob insists he cannot spell, but Mr. B. does not accept excuses. Instead, he determines his spelling level, assigns appropriate words, and teaches appropriate strategies for improving spelling. In addition, because Mr. B. offered ongoing support, timid Jennifer developed the poise and confidence to make a speech to her class. Mr. B.'s classes are proof that high expectations for both learning and behavior are effective keys to achievement and a well-managed class.

Looking Ahead

Middle school educators often consider classroom management to be one of the most difficult and challenging aspects of teaching. During the last decade or so, the research on effective teaching has provided insights into how teaching behaviors can affect—positively or negatively—student behaviors. The challenge now is to develop personal practices that have a positive effect on student behavior and overall classroom management. ■

Motivate the unmotivated
with Scientific Discrepant Events

by Emmett L. Wright, Ph.D.

Emmett L. Wright *received his Ph.D. in Academic Curriculum, Science Education, and Environmental Biology from Pennsylvania State University. He has 25 years' experience teaching science at the junior high school, senior high school, and college level, and he is the author of numerous articles. His current research interests include problem solving, misconceptions and scientific discrepant events, preservice teacher education, and curriculum development and evaluation. From 1998 to 2000 he was Chair of the National Science Teachers Association International Committee. Currently, he is a professor of science education, environmental education and curriculum at Kansas State University, in Manhattan, Kansas.*

Almost any science teacher can relate to the frustration of trying to teach an unmotivated student. How do you involve a student who, for one reason or another, just doesn't like science? One way to engage students is to use ideas and activities that are based on students' prior knowledge. This not only provides all students with familiar entry into the new material but also helps motivate disengaged students by making learning relevant to their own lives.

For example, when a teacher asks, "Why does a shower curtain move toward you when you take a shower?" or "Are peanuts really nuts?", students are presented with questions about familiar topics and phenomena. Even students who normally don't care about science will form an opinion about these questions because they have personal knowledge about the subjects. Regardless of whether their opinion is based on scientific information they have previously learned, on first-hand observations, or on intuition, students will consider the question and try to explain their answer.

Their awakened curiosity about the world around them can become a powerful springboard for learning. Once students begin to understand that science is part of their everyday lives, learning about science becomes relevant. And once this crucial step has been taken, students can begin the lifelong process of self-education.

Students' awakened curiosity about the world around them can become a powerful springboard for learning!

Discrepant Events Awaken Curiosity

A discrepant scientific event is a surprising occurrence—such as corn growing faster in the dark than in the light—that challenges learners' preconceptions. Because they at first appear to be nonacademic in nature and frequently differ from what is expected, these tidbits can stir the interest of even the chronically disinterested student. The moment a student proclaims, "No, that just isn't possible; it can't work that way," a window of opportunity opens for the student to gain a better understanding of his or her world.

For example, imagine a teacher asks students, "If you wanted to wash a car in freezing weather, would you use cold water or hot water to slow the freezing process?" Most students would choose hot water to wash the car. When students are then asked to explain their answer, they might say that hot water will take longer to freeze because the temperature needs more time to cool to the point of freezing. When the teacher proposes that hot water will actually freeze faster than cold water, this challenges the students' preconceived notions (usually based on "common sense") of the freezing process. Even if students are able to produce the correct answer because they suspect the teacher might be trying to trick them, they will remain interested because they want to know *why*. In either case, the teacher has set the stage for a scientific discovery that can stimulate even the most unmotivated students into eager and active participation.

The moment a student proclaims, "No, that just isn't possible; it can't work that way," a window of opportunity opens for the student to gain a better understanding of his or her world.

By using a discrepant event, a teacher can assess students' prior knowledge of a subject area by asking simple questions and holding relaxed discussions. These discussions can pique students' interest and generate even more questions about the subject. From this point, students can develop and test hypotheses by designing their own lab or performing a predesigned lab, such as the "Freeze to Believe" example that follows. The "Freeze to Believe" lab is an in-class demonstration that encourages student participation in the learning process.

Freeze to Believe

Exploration

Tell students that to delay the freezing of water, people who live in cold climates wash their cars with cold water instead of hot water. Ask them, "Does cold water actually delay freezing?" The majority of students will say no. In order to find out the truth for themselves, students can complete the following experiment:

Explain to students that 3 hours and 45 minutes ago you filled two stainless steel containers with 400 mL of cold water. After labeling one container "Hot," you heated it to 60°C. The second container was labeled "Cold." Explain that you placed both containers in a freezer at the same time. Next, ask students to write down which container they believe will begin to freeze first, and have them explain their reasoning. Most students will answer that the water in the Cold container will begin to freeze first because its temperature is closer to the freezing point.

Draw up a time chart that has students' names listed at 10-minute intervals for the first hour of class (this can be done in pairs). The chart should note the time you placed the containers in the freezer. Present the first pair of students with the time chart and a timer set for 10 minutes. When the alarm sounds, the first pair of students will go to the freezer and check the containers. They will record their observations on the data sheet, reset the timer, and pass the materials to the next pair of students.

Students will be checking the containers during the fifth hour that the containers are in the freezer. During this time, the Hot container will begin to show signs of freezing. Once this occurs, ask students to write down why they think this happened.

Concept Introduction

The water in the container labeled "Hot" transfers its thermal energy to the stainless-steel container faster because the temperature difference between the water and the freezer is greater. In addition, molecules in hot water have more kinetic energy than those in cold water do. As a result, hot water transfers energy to the surrounding air more quickly than cold water does. Finally, hot water evaporates more quickly than cold water does. Thus, the container labeled "Hot" has less water to be cooled. Although the amount of evaporation is very small in the container, it would be much higher if one were washing a car. Each of these points helps to explain why warm water will always reach the freezing point faster than cold water will.

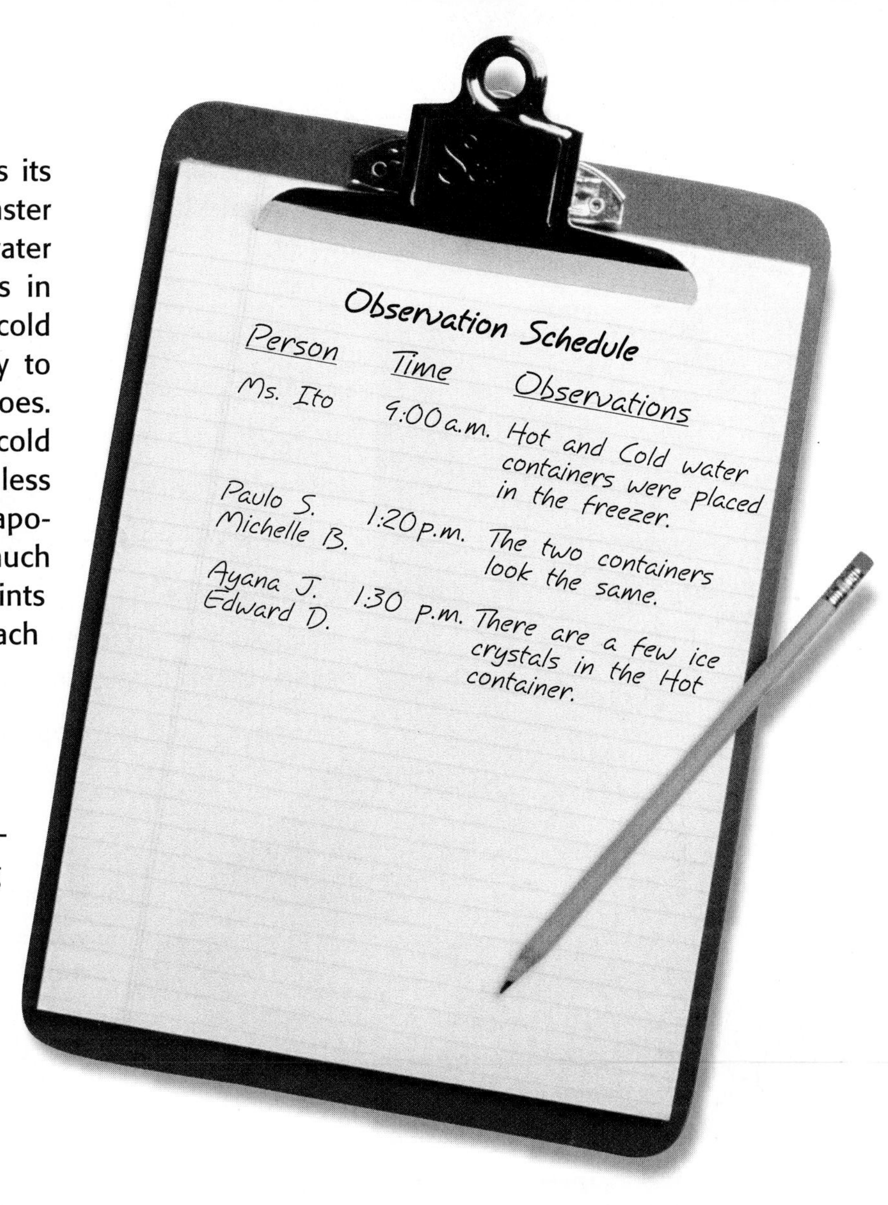

Application

Students may apply these principles by doing a project or additional research to answer the following questions: How do ice cubes cool a drink? Do the ice cubes absorb or release energy? If you were to perform the same experiment with completely insulated containers, would the hot water still freeze faster than the cold water?

Other Examples of Scientific Discrepant Events

The following are additional examples of scientific discrepant events. Some of these can be used for class discussion, while others can be adapted as hands-on demonstrations or labs.

Sink or Swim Scenerio

True or false?

When a largemouth bass *(Micropteras salmoides)* takes air into its swim bladder from the gills, the fish rises in the water. When it releases air from the swim bladder, it sinks.

Students will likely answer that this is true; however, it is actually false because the opposite occurs. When air is taken in, a largemouth bass sinks; when it releases air, it rises.

The appropriate equation for this question is:

$$D = M/V$$

(D = density, M = mass, V = volume)

When the fish takes air into its swim bladder, the fish's density, or specific gravity, increases to above 1. The air weighs more than the vacuum created when it is released. Since the specific gravity of fresh water is about 1, the fish sinks. Thus, the fish is able to sink, rise, or suspend itself by changing its density.

A Sugar Fire

Yes or no?

A cube of sugar will not ignite from a lit match. Do you think the cube will ignite if you sprinkle ashes on it first?

Students will likely answer that the cube will not ignite. However, the ashes act as a catalyst, which causes the sugar cube to ignite. This example could be used to teach about catalysts and their functions in biochemical systems.

Have Some Ground Nuts

True or false? Peanuts are nuts.

Students will likely answer that peanuts are certainly nuts. In fact, peanuts are a legume that grows in the ground. Most nuts grow on trees. In India and other tropical countries, peanuts are called ground nuts.

Does It All Taste the Same?

Yes or no?

Can all people detect the same flavors?

Students will probably answer that this is true. Scientists know of four different tastes—sweet, sour, salty, and bitter—that are recognized by the taste buds on different areas of the tongue. Some people, however, have "taste blindness" and cannot taste certain bitter compounds. PTC (phenylthiocarbamide), sodium benzoate, and thiourea are all chemicals that these people cannot taste. Whether or not people can taste these compounds is determined by heredity.

A Comet's Tail

In what direction does a comet's tail point?

Students will probably answer that a comet's tail points away from the direction it is traveling. For example, if the comet is traveling toward Earth, its tail points away from Earth.

Actually, a comet's tail always points away from the sun, no matter what direction it is traveling. The pressure from solar radiation and solar winds cause the comet's tail to point away from the sun. Students can model this effect by holding a ball with streamers attached to it in front of a fan.

Making Science Relevant

One of the best ways of motivating unmotivated students is by engaging them in learning that is relevant to their lives. This means starting with information and subjects which students already have experience in. Students' curiosity can then be captured by introducing discrepant events into the discussion. Discrepant scientific events present students with surprising facts about familiar topics. As students try to make sense of discrepant events through investigation and analysis, they become curious about the phenomena, find themselves engaged in their own education, and feel motivated to keep learning. ■

Ensuring Girls' Success in Science

by Mary Sandy

***Mary Sandy** is the director of the Virginia Space Grant Consortium, a coalition that seeks to improve math, science, technology and engineering education for all students. Sandy has co-authored a number of studies and publications on gender-equity in math and science education, and has served on the Virginia Governor's Advisory Committee on Girls' Education. She has also worked as principal investigator on grant-funded projects related to gender-equity in education.*

In the 1980s the United States awakened to the reality that our nation could ill-afford not to encourage girls to pursue studies and careers in science, mathematics, engineering, and technology (SMET). In the late twentieth century, pure economics and work force demands began to drive a national movement to stimulate girls' interest in these fields. Early initiatives by the U.S. Department of Education and the National Science Foundation assumed that simply making girls aware of SMET career opportunities and their potential rewards, as well as guiding them to the appropriate coursework, would solve the problem. The impact of this approach was marginal.

Later, in the 1990s, educational research continued to delve into the issue. An American Association of University Women (AAUW) report presented compelling evidence that "girls are not receiving the same quality or even quantity of education as their brothers."[1] David and Myra Sadker further clarified the environmental, social, and pedagogical factors that needed to be addressed in order to help girls develop their full potential in SMET fields.[2]

Girls and boys enter school with what appears to be equality of interest and performance in mathematics and science. Elementary school performance

[1] AAUW Educational Foundation, 1992, from "Forward." (Complete references can be found in *Section III: Continuing the Discussion,* page 141.)

[2] Sadker and Sadker, 1994.

generally continues this way, though girls tend to become more passive in the classroom by second grade. By the middle school years, which bring the onset of puberty and the allowance for elective class choices, girls' interest and performance in math and science begins to steadily decline. Enrollment in advanced mathematics and science courses drops significantly during the high school years. By failing to take the foundation courses upon which SMET college studies build, girls limit their range of career possibilities and technological survival skills needed in an information-based and technology-driven society. Hence they perpetuate an alarming trend of setting themselves up for lower-paying professions.

In order to prevent further perpetuation of these trends, educators must take measures to encourage girls in the math and sciences before they reach college. Only by fostering girls' self confidence in technical fields, by teaching with gender equity in SMET courses, and by offering extracurricular enrichment programs that make girls aware of their science opportunities can teachers provide a foundation upon which girls can grow into successful mathematicians, scientists, and engineers.

Girls and boys enter school with what appears to be equality of interest and performance in mathematics and science.

Fostering Success and Self-Esteem

We socialize our children in such a way that the "helpless female" myth still flourishes in our society. Though it may be unintentional, parents still treat their daughters and sons very differently, holding the different genders to different standards. Whereas boys are more often encouraged to take risks and allowed to learn from their mistakes, girls are more often protected and taught to be risk-averse. One of the hallmarks of success in engineering and science is tenacity. This means that we must help girls to thrive in problem-solving environments.

Ways to promote success and self-esteem

1. Award praise and positive feedback equally and fairly.
2. Use language that nurtures and encourages students.
3. Make positive statements about math and science.
4. Stress that girls and women can succeed in SMET fields.
5. Keep expectations high for all students.

Building Personal Toolboxes For female students to overcome societal limitations and to succeed in SMET studies and careers, they will need a set of personal success tools. Parents, educators, and society must help girls build and fill their personal toolboxes. We can begin by instilling in them the ability to take risks and to make mistakes and learn from them without personalizing their mistakes. We can also lend support by helping girls develop self-advocacy skills and build a durable sense of self-worth.

Sending the Right Message The messages teachers, parents, and society send to students affect their perceptions of themselves and their abilities. In school and at home it is important to build confidence in girls. To do that, we should encourage girls to take risks and to have faith in their abilities to solve problems and act independently. Praise and positive feedback are essential building blocks for self-esteem. Teachers tend to praise girls and boys differently and for different reasons, sending powerful messages about ability. We must seek to offer words that nurture and encourage students versus language that limits them. It is important to be positive in statements we make about math and science and to stress that girls can be successful in these fields. Girls far too often get the message that "math is hard" or "scientists are nerds." Educators and parents need to give plenty of positive reinforcement and keep expectations high for all students. Such reinforcement becomes critical as girls enter puberty, when they are most vulnerable to peer pressure and most likely to retreat from challenges that put them in competition with the opposite sex, in and out of the classroom.

Teaching with Gender-Equity

The Sadkers found that girls and boys receive very different treatment in school, resulting in very different educational experiences.[3] Boys and girls are generally held to different behavioral expectations in the classroom. While girls are more often held to higher standards for neatness and good behavior, a societal attitude of "boys will be boys" allows male students to push behavioral limits. They more often get by with breaking rules, shouting out answers in class, and demanding the teacher's attention. When girls speak out loudly, they are often reprimanded for their behavior. As a result, we see girls dropping out of the competition for classroom attention.

Equal Participation One strategy for ensuring participation by all students is to put their names on index cards (each stack should have two or three cards with the same student's name). The teacher shuffles the cards at the beginning of each class and uses these cards to call on students in a random fashion to maximize participation. Teachers also need to be sure that they are asking the same kinds of questions of all students (probing, analytical, evaluative, synthesizing). All students should be held to the same disciplinary standards regardless of gender, and the teacher should strive not to make comparisons of students that are gender based.

It is important

that the classroom climate be a postivite one that supports all students in their quest for knowledge...

The Classroom Climate It is important that the classroom climate be a positive one that supports all students in their quest for knowledge and that engages girls as vigorously as boys. The astute teacher will recognize the different behavioral tendencies of boys and girls and will apply strategies that nurture the learning of all students. Real-world relevance makes mathematics and science more meaningful for all students, but it seems to be an essential element for girls. Because girls are socialized to place more emphasis on societal and humanistic concerns, real-world relevance gives them a greater opportunity to approach topics from a familiar point of view. The textbooks used in class must reflect the accomplishments of women scientists and technology policy leaders and must show the diversity that exists in the world in which we live. Also, teachers should ensure that girls play active roles in class activities and that classroom materials have both girls and boys as leaders or heroes.

[3] Sadker and Sadker, 1994.

Group Learning While competitive learning situations tend to be most effective for white male students, cooperative learning situations and effectively designed research or project teams can work well for all students. The following, for example, is one effective strategy: The class is divided into teams. Students from each team meet in specialty groups in order to learn information essential to performing the research or undertaking the project. When students subsequently rejoin their team to continue the activity, each student is a resident expert in his or her assigned field. This approach requires that all members contribute to the team effort. Teachers should ensure that all students equitably assume a variety of roles, which ensures equitable learning.

Building a Gender-Fair Learning Environment

Try these strategies for creating a gender-fair learning environment:

1. Ensure participation by all students, asking the same kinds of questions of all students.
2. Hold all students to the same academic disciplinary standards, regardless of gender.
3. Avoid making gender-based comparisons of students.
4. Monitor classroom environments to be sure that they are gender-fair.
5. Review textbooks to be sure that they reflect the accomplishments of women in science and technology.

Skill Building You may want to include in your lesson plans activities that offer many approaches to tasks and have many right answers. These activities allow students to be successful at their own levels and permit teachers to reward a variety of problem-solving skills and approaches. Tasks involving hypothesizing, estimating, and testing give girls important skills reinforcement. Foster curiosity in all students by letting them tinker with machines, take things apart, and figure out how things work. *How to Encourage Girls in Math and Science*[4] identifies four skills areas that research and standardized tests have shown to be problematic for girls. To ensure later success in SMET courses, the authors encourage teachers and parents to reinforce girls' skills in the following areas: spatial visualization, problem solving, logical reasoning, and scientific investigations. The book offers a wealth of strategies for hands-on, positive approaches to building these skills for girls.

Professional Development Most teachers want to be fair to girls; however, inequitable practices are so insidious that focused professional development is needed to raise awareness of the problem. Even the most seasoned teachers have found that they have practices that are not gender equitable. A strong commitment to change, coupled with a program of peer and self assessment, will further educators' goals of ensuring a gender-fair learning environment. Teachers may choose to videotape their classes as a means of monitoring teacher-student interactions. Assessment tools such as those offered in *A Guide to Gender Fair Education in Science and Mathematics* can be used to evaluate classroom environments and educational materials for gender fairness.[5]

---

[4] Skolnick, et al., 1982.

[5] Burger and Sandy, 1998.

The Role of Mentors and Extracurricular Enrichment

After-school computer and science programs that nurture and encourage girls' interest in SMET disciplines have been successful, especially at the middle school level. For example, computer clubs for girls allow girls to develop essential computer skills in a female-friendly, noncompetitive environment. Your faculty may also choose to host an annual mathematics, science, and technology conference, complete with guest speakers who are successful women in math and science fields. Hosting such a conference can raise parent and student awareness of academic and career opportunities and can convey important information about how students can take advantage of those opportunities.

Exposure to professional women who are self-confident and willing to encourage young people can positively affect girls' choices and their faith in themselves. Having the guidance of mentors—trusted advisors who can serve as role models—has proven particularly effective for girls. To provide this type of contact, educators can look to the community and avail themselves as mentors and role models. Professional societies, the AAUW, the National Science Foundation, and other organizations often support and encourage such enrichment programs.

Our Influence Counts

Educators not only play a crucial role in SMET education but also influence girls' attitudes and choices regarding studies and careers. It is essential that educators examine their own attitudes and practices in order to ensure the best possible education in SMET for all students. In order to reach that goal, they must first understand and then address the special needs of girls. Math and science teachers have the ability to change girls' futures by presenting material in a female-friendly way and using methodologies that work for girls as well as boys. ■

Teaching Science to Students with Limited English Proficiency

by Berty Segal Cook

Berty Segal Cook *has 12 years of classroom experience with Limited English Proficiency (LEP) and foreign-language students. For the past 18 years, she has served as a consultant and teacher-trainer, giving seminars and lecturing on teaching LEP students for school districts, colleges, and professional organizations across the United States and abroad. Cook is the author of* The Practical Guide for the Bilingual Classroom *and* Teaching English Through Action, *both of which have been translated into several languages.*

Many middle school students whose native language is not English may already have an intermediate level of English proficiency. These students may exhibit strong skills in speaking, reading, writing, and comprehending English, or a combination of these skills. Yet when learning new science material, these students may lack the confidence and tools they need to learn the material successfully. Fortunately, there are some instructional strategies you can employ that can help. The strategies presented here are geared toward the LEP students in your classroom who possess an intermediate level of English proficiency, but the strategies will benefit the other students in your class as well.

Assessing Prior Knowledge

Before approaching new material with your LEP students, find out how much they already know. At the middle school level, many LEP students will have had considerable exposure to English. They may have a solid background in listening comprehension and some experience in speaking, reading, or writing in English. In addition, some may have had experience with the science content but may have learned science in their native language.

To fully assess students' knowledge, you will need the help of another teacher or an aide. If you do not speak your LEP students' native language(s), try to enlist a teacher or aide who does. Begin by introducing a new topic to the class. If your assistant is familiar with the native language(s), he or she should focus on the LEP students while you work with the rest of the class. Begin with each group discussing the topic in a fun and informal way. In both groups, have students share their prior knowledge on the subject, including vocabulary, definitions, and concepts. Encourage students to help one another clarify concepts and definitions. Freely offer the correct vocabulary terms if a student is able to explain a concept but cannot remember the terminology.

This sort of preliminary discussion allows LEP students to draw from their experience and gives them the opportunity to contribute to class discussions. It also gives LEP students a sense that what they have already learned may be helpful in the class. During the discussion you may find that you need only review with them new English vocabulary for concepts they have already learned. Students therefore do not have to start off in their new educational environment with the feeling that they are being forced to start over. This can be a critical confidence builder for LEP students.

Taking on New Content

Before opening the textbook, keep in mind that because these students possess a limited English proficiency, you will need to provide them with additional learning tools. Help them get started by developing some concepts *before* beginning a new book or unit. Following are some effective strategies for teaching new content to LEP students.

Effective Strategies for New Content

- Give an overview
- Explain your expectations
- Use hands-on activities
- Use discussion groups

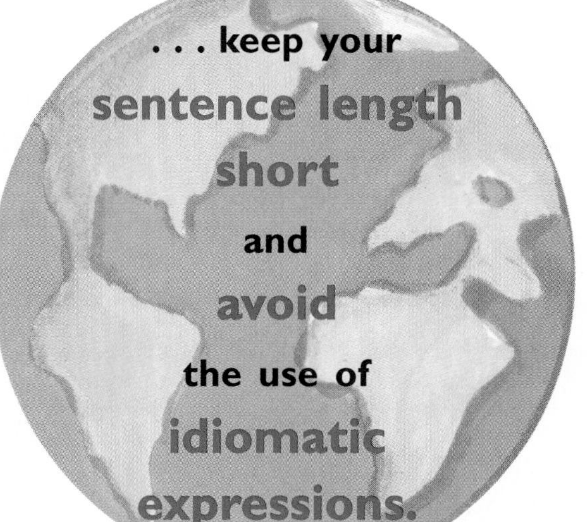

Give an Overview

Provide your students with an overview of the chapter or unit. Use as many visual aids, transparencies, and models as possible while you are discussing the new material. It is important to identify the vocabulary words and main concepts with your students while using the visuals. This builds the students' familiarity with the concepts to be learned, gives them a head start with the vocabulary to be covered, and increases their confidence that they will be able to follow the lessons. It is also worthwhile to include other contextual clues during this overview. Specific gestures, facial expressions, or even acting out the meanings of words will help some students better grasp the meaning of new vocabulary terms.

It is important to reduce the pace at which you convey new information. This means speaking slower and enunciating clearly. Be mindful of your word choice when explaining concepts so that you are not presenting students with new vocabulary that is unnecessary for the lesson and for your goals. It is also a good idea to keep your sentence length short and to avoid the use of idiomatic expressions.

Students who have less than an intermediate level of English proficiency should preview concepts in their primary language. This is a three-step process that can be summarized as follows:

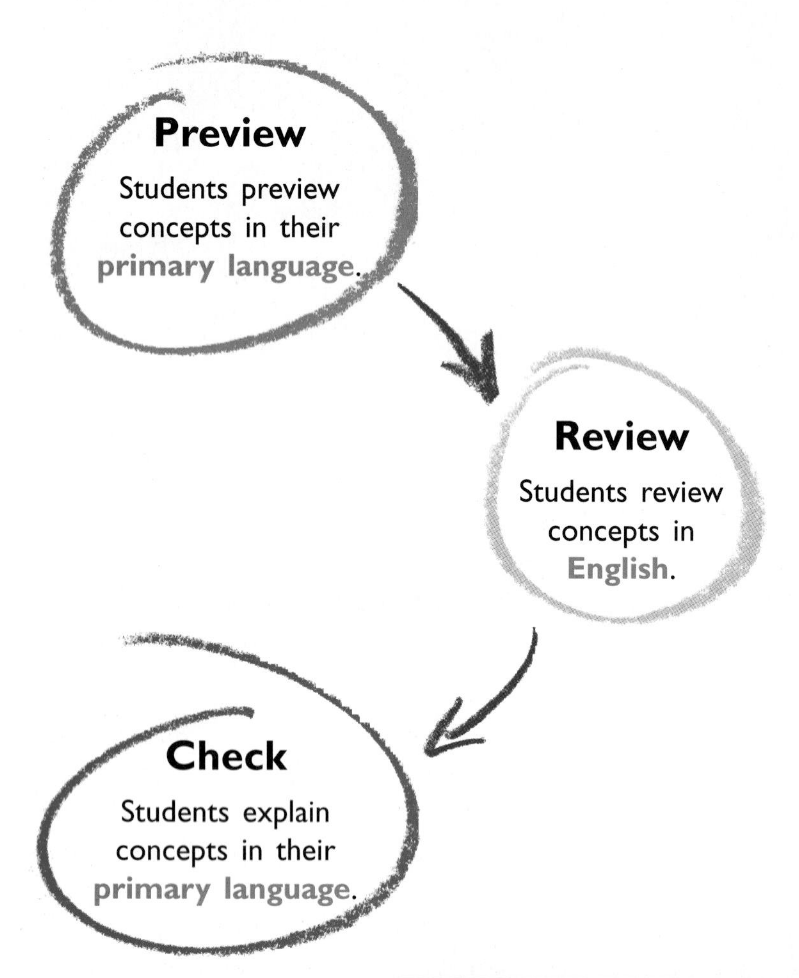

When possible, preview new concepts with the students in their native language. Your aide can do this during class time while you give a similar overview in English to the rest of the class. Once the preview is completed, you should review the concepts with the LEP students in English while the aide reviews the same concepts with the remainder of the class. Finally, if possible, the bilingual instructor will check the LEP students' understanding of the concepts in English by meeting with the LEP students a final time and speaking to them in their native language. The aide can go over the concepts and then ask the LEP students to explain those concepts in their own words and in their native language. During this time you should perform a similar check with the remaining students.

Explain Your Expectations

After the overview, give your students a detailed list of your expectations. This list might include the amount of time you estimate it will take to cover the material, the vocabulary and concepts they should be familiar with at the conclusion of the lesson, the types of quizzes or tests they can expect, and the labs or activities they may be doing. With an overview of the material and a clear sense of what is expected from them, your students should be able to approach the material with confidence. From this point forward, you should work with the entire class.

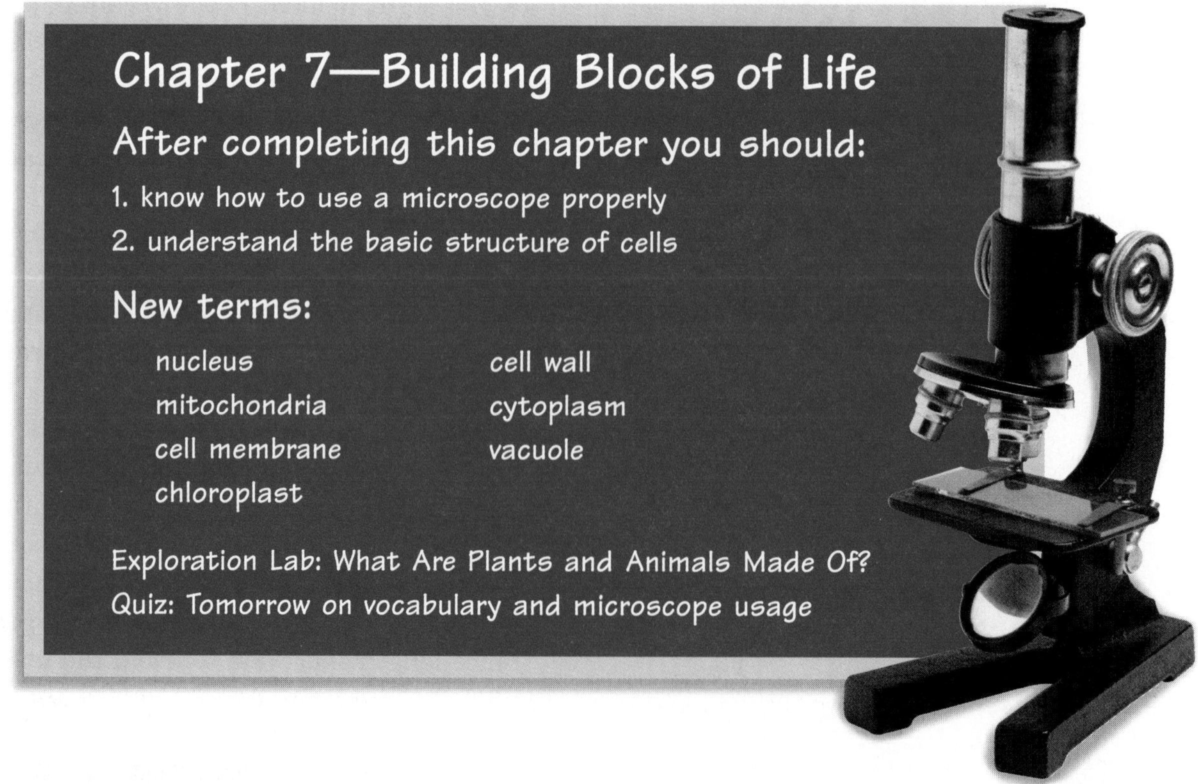

Use Hands-on Activities

Once you have given your students the confidence that they can handle the new material, introduce the new material by engaging the students in experiments and activities that present the material in a hands-on manner. You can also perform demonstrations related to the material. Intersperse or follow these activities with visual aids, such as photos, transparencies, or charts. While using visual aids, point out the essential vocabulary and concepts you presented in the overview.

At this point, it is a good practice to make frequent checks for understanding among all the students in your class. Your LEP students may need you to clarify concepts and definitions and would benefit from being asked a variety of questions regarding their comprehension.

Use Discussion Groups

Set aside sufficient time to encourage your students to talk about what they have observed in the demonstration or hands-on experience. Encourage the students to discuss their ideas and results with you. At this point in the lesson, students with an intermediate level of English proficiency should be able to discuss the topic alongside their native-English-speaking peers. Their level of English should also be advanced enough that they are capable of explaining a concept or term without using the exact term necessary because they may not know it or remember it. By explaining terms and concepts, LEP students can get their point across and will benefit from having the proper English term provided for them. It serves as a reinforcement because the more opportunities LEP students have to interact with new material, the more likely they will master it. This is true for all students, however, and as a result, this activity will help your non-LEP students solidify the new material in their minds as well.

By writing down and then reading aloud the students' own words, you give considerable value to what the students have said and a great boost to their self-esteem.

During the discussion, write down what your LEP students say, including any grammatical errors. Review each student's impressions individually, and address grammatical and conceptual corrections as you go. This approach establishes a pattern of listening and speaking activities before students are asked to read and write on the topics. It is especially helpful in encouraging LEP students who are unsure of their abilities. By writing down and then reading aloud the students' own words, you give considerable value to what the students have said and a great boost to their self-esteem. This approach also acknowledges and clarifies what the students understand, reviews the concepts they are learning, and prepares them for reading about the material in the textbook. At this point, all your students are better prepared to interact with the material through reading.

Using the Textbook

Provide Focus Questions

As you progress through the lesson, it will be helpful to you and to your LEP students if you provide focus questions. These questions should act as a guide to students, giving them an idea of what to look for as they read. Focus questions can refer back to the overview you presented earlier and to the demonstrations or hands-on experiences. This way, the students can connect the previewed concepts with the materials in the textbook. One method for sharing focus questions with students is to write the focus questions on a chalkboard that students can refer to as they read the text. Ask students to read only those focus questions for the section they are about to read and to then write out their answers to the questions when they have completed the reading, before moving on to the next section.

Sample Focus Questions

How have the Earth's oceans changed over time?
What is a cell, and where are cells found?
What is matter?
How can you determine if some rocks and fossils are older than others?
What is the difference between a rock and a mineral?
How are compounds and mixtures different?
What does it mean to work?
Why don't all humans look exactly alike?
What do plants use flowers and fruits for?

Provide Graphic Organizers

Provide your students with a visual strategy for learning by using outlines, papers with headings and subheadings, compare-and-contrast columns, and five-inquiry columns—who, what, where, when, and how. These are effective tools to help your students approach the material in an organized manner. Later, these hand-outs can be used as study guides in preparation for the test.

Cells: The Basic Units of Life

I. Cells: Starting Out Small
II. Tissues: Cells Working in Teams
III. Organs: Teams Working Together
IV. Organ Systems: A Great Combination
V. Organisms: Independent Living
VI. The Big Picture

Learning New Vocabulary

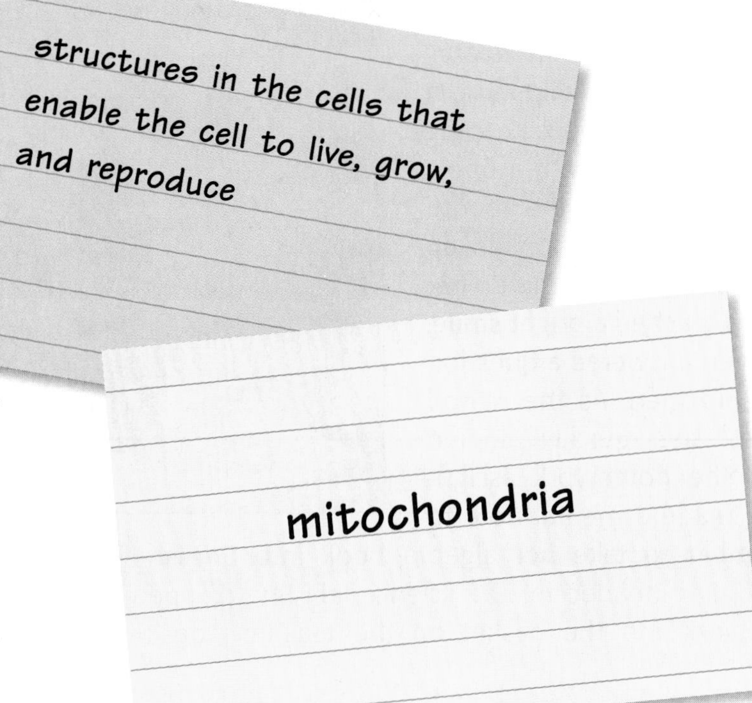

As you move through the text, point out the vocabulary and labels that you presented orally in the earlier steps. It is critical that the vocabulary be presented and learned in context, not in lists. After reviewing the terms in context, have students create a flashcard for each word, putting the word on one side and the definition on the reverse side. When appropriate, encourage your students to add a drawing or small diagram to the definition side of the card. This use of visuals is an extremely effective learning tool for LEP students because the right side of the brain, which controls memory, thinks in pictures. Students can use these flash cards later in oral pair work and to prepare for tests. You may also want to have students maintain a science vocabulary book with a running list of words and phrases. Students will be able to use this as a quick reference. Again, all vocabulary should be presented in context as well as in dictionary form.

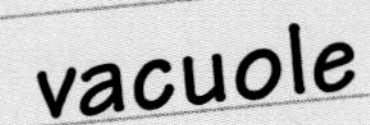

Discussing and Reading

Conduct discussions in your class, and encourage discussion while students are reading together. You may even want to use oral "read-around groups."

To conduct a read-around group, ask students to sit in a circle on the floor. This is intended to be a very informal and comfortable exercise in which students take turns reading aloud from the text. When students have difficulty with pronunciation, those students with greater language proficiency can help those with lesser proficiency. Throughout this activity, it is important to check your students' comprehension. Instead of asking questions that begin with "Do you understand?", ask your LEP students simple inquiry questions about the text such as, who, what, where, when, and how?

Proven Success

The strategies presented here have been proven to make a strong difference in how much LEP students learn and how much they enjoy learning. With these techniques in hand, you can teach your LEP students with confidence. I send you my best wishes for success and enjoyment in your teaching! ■

Keisha ignored the antics of her friends in nearby seats and quickly whipped out her science lab report. Keeping her work hidden with a book, she quickly wrote the report. Every so often she looked up and gave Mr. Bennett, her algebra teacher, a bright smile. She even answered a question now and then. As the period ended, she quickly copied down the homework assignment, made a mental note to remember to take her algebra book into English fifth period, and headed for the science lab. She dropped her homework into the basket on the teacher's desk and took her seat. Miss Adams glanced through the lab reports as the eighth graders completed their reading of the last section of the chapter. As usual, Keisha's work was excellent. Before beginning the class discussion, Miss Adams asked Keisha to share with the class the steps she took in creating such a thoughtful, thorough, and well-planned lab report. When she heard her teacher's request, Keisha blushed and dropped her eyes. She worried, "What can I say? I didn't plan anything! I just wrote it five minutes ago. Should I lie?"

Meeting the Needs of the Academically GIFTED

by Sally M. Lafferty, Ed.D.

Sally Lafferty *has 28 years of experience in science education, gifted education, curriculum development, and middle school administration. She is currently the supervisor of science and of the Gifted and Talented program for the Salt Lake City School District, in Salt Lake City, Utah. Lafferty is also the chair of the Middle School Division of the National Association for Gifted Children.*

Dare Keisha admit publicly that she wrote the report during math class, the period before? If she made up a different story, would another student report what she had done, exposing her lie? Should she say what she knows the teacher wants to hear, or should she tell the truth? Students for whom academic endeavors come easily struggle with these dilemmas daily. At the very moment at which they are being lauded as model students, many gifted students feel like impostors. They know they do not need to work long and hard and that they did not master challenging tasks with diligence and perseverance. Many gifted students are truly sorry that assignments are so easy for them. Others no longer care. While most gifted learners remain compliant (seemingly attentive while reading their novels hidden within their texts), others become provocative (asking questions that lead class discussions on tangents and otherwise destroying teachers' lesson plans), or even rebellious (refusing to participate in class activities or work that does not interest them). In these ways, the potential for production and performance of many gifted students is lost as they mentally "check out" of school. How can teachers make a difference for these exceptional students? How can we challenge and inspire them with engaging work without leaving the rest of the class behind?

Knowing Your Students

Teachers who embrace the concept of flexible teaching methods can make conscious modifications to their lesson plans to better meet the needs of all their students. Perhaps the first step in effectively teaching gifted students is the teacher's acceptance that all students are different. No single teaching style, lesson plan, learning expectation, assignment, or project can be appropriate for all students. Students in the same class may have widely disparate readiness levels, abilities, and interests. They are motivated by different stimuli, engaged by different tactics, and inspired by different internal and external rewards. Indeed, classroom research suggests that the disparity among the ways students learn may be greater at the middle school level than later in their education.[1] Therefore, middle school educators must be creative and flexible in their teaching methods for all their students. But before a teacher can develop or modify curriculum and instruction to meet the needs of gifted learners, he or she must first recognize the differences between the truly gifted student and the high achiever.

High Achiever or Gifted Student? High achievers are usually described as interested, attentive, and hard-working. High achievers pay attention in class to absorb all that a teacher says. Gifted learners, on the other hand, may be quite inattentive and still pass the test with ease. High achievers answer the questions asked; gifted learners ask the unanswered questions. A high achiever is content to understand an idea, complete an assignment correctly, and know the information presented. A gifted student wants to construct new ideas, theorize about them, and change an assignment into a more complex project.

Other noticeable characteristics of truly gifted students are the fast pace at which they learn new material and the vast amount of knowledge they bring into the classroom. Many are almost compulsive in their need for intellectual stimulation. In addition, gifted learners have the ability to make intuitive leaps, making sense of isolated facts by relating them to one another. Gifted students often emphasize the relevance of an activity or task and often need to understand how the activity fits into the "big picture" before they are willing to give it their full attention and effort.

Teaching Gifted Students

There are many excellent teaching strategies that will help teachers better meet the needs of gifted students while extending the learning opportunities of other students. These strategies will raise the level of instruction for all students and remove the "glass ceiling" that often prevents gifted students from working above the level of other students. The work of Carol Ann Tomlinson at the University of Virginia, much of which is specific to middle school teaching and learning, is particularly worthy of teachers' review. Her first premise is that all students need relevant content material combined with engaging instruction and challenging assignments.[2] No visitor or participant

[1] National Middle School Association, 1995. (Complete references can be found in *Section III: Continuing the Discussion*, page 141.)

[2] Tomlinson, 1994.

in a classroom should be able to identify students' learning levels simply by noting their activities and assignments. In order to engage students at different learning levels in exciting, meaningful learning experiences, Tomlinson promotes a differentiated classroom based on the principles listed in the box below.

Principles of a Differentiated Classroom

- ★ The teacher understands, appreciates, and builds on students' differences.
- ★ Assessment and instruction are inseparable.
- ★ The teacher adjusts content, process, and product in response to students' readiness, interests, and learning profiles.
- ★ All students participate in respectful work.
- ★ Students and teachers are collaborators in learning.
- ★ Goals of a differentiated classroom are maximum growth and individual success.
- ★ Flexibility is a hallmark of a differentiated classroom.

Source: Carol Ann Tomlinson, University of Virginia, 1997.

Tomlinson suggests that teachers begin units of study by selecting a concept in a core curriculum or discipline, identifying one or more relevant generalizations, and designing a tiered lesson around the generalizations. A tiered lesson is a teaching strategy in which teachers assess the readiness level (interest, prerequisite knowledge, and skill level) of their students and group them accordingly for both instruction and production.

For a given lesson or unit, the teacher can plan from three to six different exercises that focus on the same concept and generalizations but use resource materials and learning activities of increasing difficulty and complexity. Because students' readiness levels will be different for each topic, student groupings will vary as the class moves through the curriculum.

A TIERED LESSON in Action

Here's an example of a tiered lesson as it would unfold in a classroom. For an integrated science unit on classification, the teacher would choose a related generalization, such as, "People use classification to organize and structure many things." Then the teacher would pretest students and assess their readiness for tackling this topic. Based on the results of the pretest, the students would be divided into four learning groups for several days of explorations.

Group One might begin by focusing on properties of matter by reading about and experimenting with density and temperature. Their further explorations could take them from the physical properties of water through the characteristics of living and nonliving things.

Group Two, having demonstrated mastery of Group One's basic knowledge in the pretest, may be asked to describe the identifying characteristics of a variety of commonly collected objects, such as insects or leaves. After selecting a different group of organisms, the students would then arrange the organisms into groups, carefully describe their reasons for including each organism in its respective group, and develop a system that they can relate to scientific classification.

Group Three, whose members have been assessed to have a good sense of scientific classification, might select a kingdom, identify the characteristics of the organisms within it, and delve into the patterns of taxonomic classifications. As a group, students could conceive of a "new" organism and attempt to classify it scientifically.

Group Four, understanding the overall system of scientific classification of living things, can be challenged to look for similar patterns of classification in other disciplines. How is the structure of scientific taxonomy like the structure of a business or a school? How does structure evolve? How is it destroyed? Ethically speaking, can one debate the need for structure? How do different people feel about structure?

No teaching style,
lesson plan, learning expectation,
assignment, or project
can be appropriate
for all students.

Tiered lessons offer all students a match between their personal readiness level and content level, and they allow all students opportunities to challenge themselves. Ideally, the challenges are neither too easy nor too difficult, but just hard enough to be engaging without frustration or fear. Also, students gifted in math should not be expected to be equally gifted in creative writing. Likewise, learners struggling in history class may not struggle in science class. Every student is held to realistic expectations of growth and presented with appropriate challenges across the disciplines.

Common Concerns with Differentiated Classrooms

Two issues may discourage teachers from embracing the concepts of a differentiated classroom and tiered lessons. First, traditional assessment tools must be replaced with models that allow teachers to quickly assess and group students for learning. Because a differentiated classroom is regrouped often, assessment must be a constant and ongoing process, rather than a single evaluation at the end of a course of study. Skill and product rubrics must be created for this type of assessment. In a differentiated classroom, the teacher must strive to know every student well and use this knowledge constantly. This takes a considerable amount of planning, practice, and energy.

Another question many teachers have about differentiated classrooms is how the grouping of students affects the learning of the class as a whole. Some teachers may claim that classroom differentiation is just another form of ability grouping. The debate over the issue of ability grouping has raged for years, and with good reason. The fact that gifted learners benefit by peer association is not contested, but the question of the extent to which teachers should allow groups of gifted students to work together is still hotly contested.[3] The differentiation of a classroom is not a method of ability grouping. Instead of being grouped according to an assessment of overall ability, the differentiated class is grouped by individual readiness for specific topics and concepts. Classroom differentiation allows students to flourish in different content areas while remaining integrated.

[3] Kulic, J.A., and L.C. Kulic, 1987.

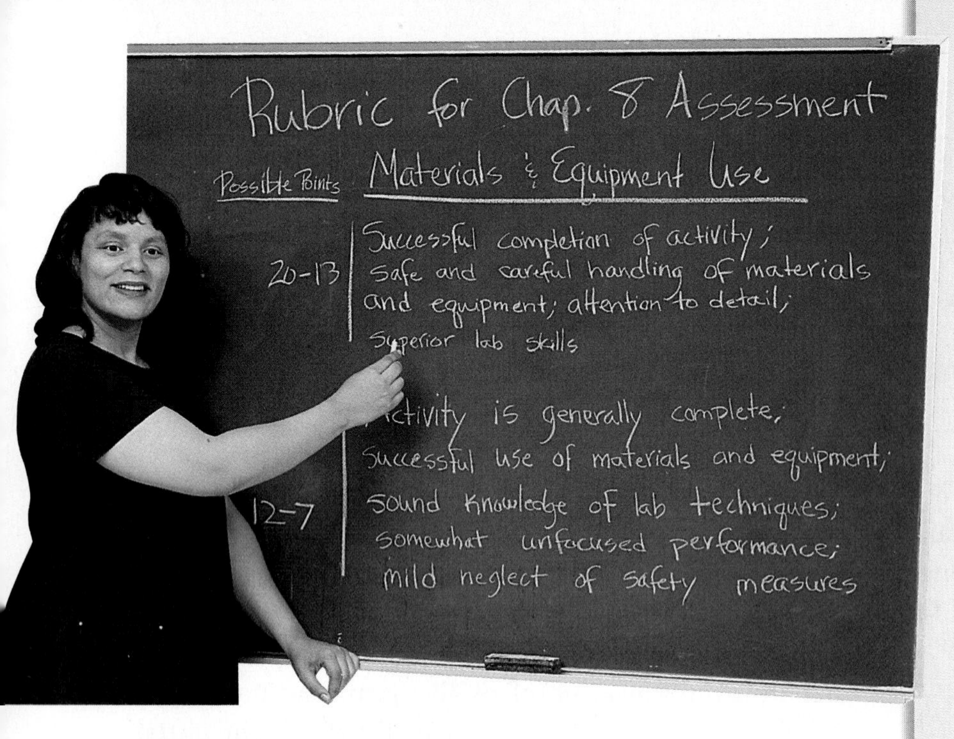

More Strategies for Gifted Students

There are many other teaching strategies that may be successfully employed in differentiated classrooms. For example, supplying your student with more than one text on a topic presents different perspectives and demonstrates that there is not always one right answer or view. Matching questions to specific students helps each student to expand his or her comfort level and explore risk-taking challenges. Self-assessment through the use of rubrics teaches students to seek personal improvement and maximum growth and to hone judgment and evaluation skills. Graduated rubrics—rubrics developed by the students and teachers and modified as the students develop greater mastery and skill—demonstrate student maturity and progress.

Moving Toward Depth and Complexity Sandra Kaplan, associate professor of learning and instruction at the University of Southern California, suggests that teachers move students, especially gifted students, into greater depth and complexity, guiding them toward academic excellence and scholarliness.[4] Kaplan recommends eight ways that teachers can guide students to a greater depth of understanding. These strategies, while particularly helpful in lifting the glass ceiling for gifted learners, will help all students excel academically.

[4] Kaplan, S., keynote address, 1998.

8 Ways to Greater Understanding

1. **Use the Language of the Discipline** Open the door to the discipline by modeling the use of the real vocabulary and specific terms.
2. **Embellish with Details** Identify the attributes, gather the facts, and describe specific features and characteristics. Use details to elaborate on the larger questions.
3. **Determine Patterns** Examine the body of knowledge for patterns, events, elements, and ideas, especially those that are repeated over time. Consider the sequence of events, and encourage the use of current knowledge to make predictions. Teach young scientists to "read between the lines," searching for the logic behind the words. Hypothesize, prove, and defend.
4. **Investigate Trends** Look for connections outside the topic. Note ongoing factors that influence or contribute to the discipline. Search for the forces that are shaping the discipline. Challenge learners to think about the political, ethical, or social effect of scientific topics.
5. **Ponder the Unanswered Questions** Identify what is known, and encourage discussion about what is not known. Identify incompleteness in disciplines, and encourage students to look for inconsistencies. Theorize about the reasons for incompleteness.
6. **Hypothesize About the Rules** Experiment with structure. Discuss the stated and unstated assumptions within a discipline. Depart from the standard organization. Build new structures and classifications of information.
7. **Explore the Ethics** Tackle the tough issues in the discipline. Identify and discuss dilemmas and controversies. Consider the impact on people, and discuss how scientists should use what is known.
8. **Identify the Big Ideas** State the founding theories and principles of each discipline. Find the connections and the interrelationships that give meaning to what students learn. Discuss what people hope to achieve by carrying out scientific investigations.

Another method a teacher may use to reach gifted learners is to add greater complexity to lessons designed for the entire class. Kaplan recommends three broad strategies for doing so.[5]

To Add Complexity:

1. Explore relationships over time.
2. Foster multiple perspectives.
3. Encourage relationships among, between, and within the disciplines.

Common Myths

There are many myths prevalent in school corridors and classrooms about gifted students.[6] One is that students who are gifted in one academic discipline are gifted in all disciplines. Another common myth is that gifted students can make it on their own, with limited support and guidance from teachers and administrators. In truth, like other students, gifted students have areas of strength and weakness.[7] And gifted students, such as Adam (see the box below), may need specialized support in areas besides academics.

Some gifted students may also be perfectionists. These students, like all students, face challenges inside and outside of the classroom. The perfectionist may have difficulty completing even the simplest assignment on time because he or she is struggling for the perfect report or project. Other gifted students become outwardly proud of their abilities, even to the point of arrogance, and cannot make friends. Still others, long bored with the classroom routine, lose the motivation to succeed in school. As teachers become more familiar with gifted students, they learn that gifted students are in many ways like others in their age group, suffering many of the same emotional and academic challenges.

In Summary

Every student, including the gifted student, needs and deserves to be challenged by a curriculum matched to his or her abilities. As teachers, it is our job to make this learning environment a reality for each of them. Tiered lessons and assessments in a differentiated classroom are some of the tools that teachers can use to help all of their students to succeed academically. Likewise, teachers must also ensure that all students receive the emotional support and guidance that make learning possible. With this support, all of our students may reach their highest potential in and out of the classroom. ■

[5] Ibid.

[6] Clickenbeard, P.R., 1991.

[7] George, P., 1995.

Adam trembled as he walked into his science class, his notes and charts ready. Today he had to give his report titled "The Structure and Function of Cells in Living Organisms." Adam had wanted to bring a slide presentation he was working on at home about the tomatoes he was growing, or maybe turn in his creative writing piece called "To Be or Not To Be Osmosed." But his teacher instructed each student "to give an oral report on one important aspect of Chapter 7, using at least one chart mounted on a sheet of colored poster board." On the way to school, Adam realized that his chart was on white paper. Would he lose points for that? Would he even remember to use the chart in his presentation? Adam's forehead was covered in beads of sweat. Why did he have to present orally? Adam had never spoken in front of a group before, and he was scared. "Adam, for goodness sake!" Miss Brooks had responded when he asked to do the assignment differently. "You will have no problem with this. You do everything so well." Why couldn't Miss Brooks see that he was too scared to get up in front of everyone? Why does she think that everything is so easy for him?

Making Hands-on Doable

by John G. Upham

John G. Upham *received his master of science in education from the State University of New York. For the last 26 years, he has taught middle school science at Norwood-Norfolk Central School District, in New York.*

It is widely accepted that hands-on experiences are a vital part of a child's education. This is especially true in the sciences. However, the vision of students engaged in meaningful scientific exploration in a modern lab environment can be difficult to translate into reality. While lab equipment and supplies become more expensive with each passing year, classroom budgets also seem to be continually shrinking. So how do we teach hands-on science to students without breaking our classroom budget on expensive lab equipment and supplies?

There are many strategies you can use to provide an educational, fun, exciting, and most of all, inexpensive lab-based science program for your students. All it takes is flexibility, a willingness to plan ahead, and a little creativity.

Preparing the Classroom

In this day of overcrowded schools and tight budgets, many science teachers find themselves teaching in classrooms that were not designed for science classes. In many cases, this means a lack of the most basic laboratory needs, such as faucets, sinks, and lab tables.

Among these obstacles to an effective lab program, perhaps the easiest to overcome is the lack of a water source. An inexpensive solution is the use of plastic dishpans and milk jugs for carrying and storing water. Large plastic dishpans are available at most grocery and discount stores. They are durable and will last for years. These bins can also double as storage containers, and they stack conveniently when not in use. One-gallon milk jugs can be used to transport water from the faucet, and they also make a portable water supply for each lab setup. I have found that one jug per lab group is sufficient for almost any lab in the middle school curriculum.

The design of the average middle school classroom usually means no lab tables. This is a serious problem because normal student desks are too small. Pushing desks together is a poor substitute for lab tables because they tend to separate easily, creating an unsafe lab area. A quick look at the price of suitable tables from a classroom furniture catalog is enough to cause heart palpitations.

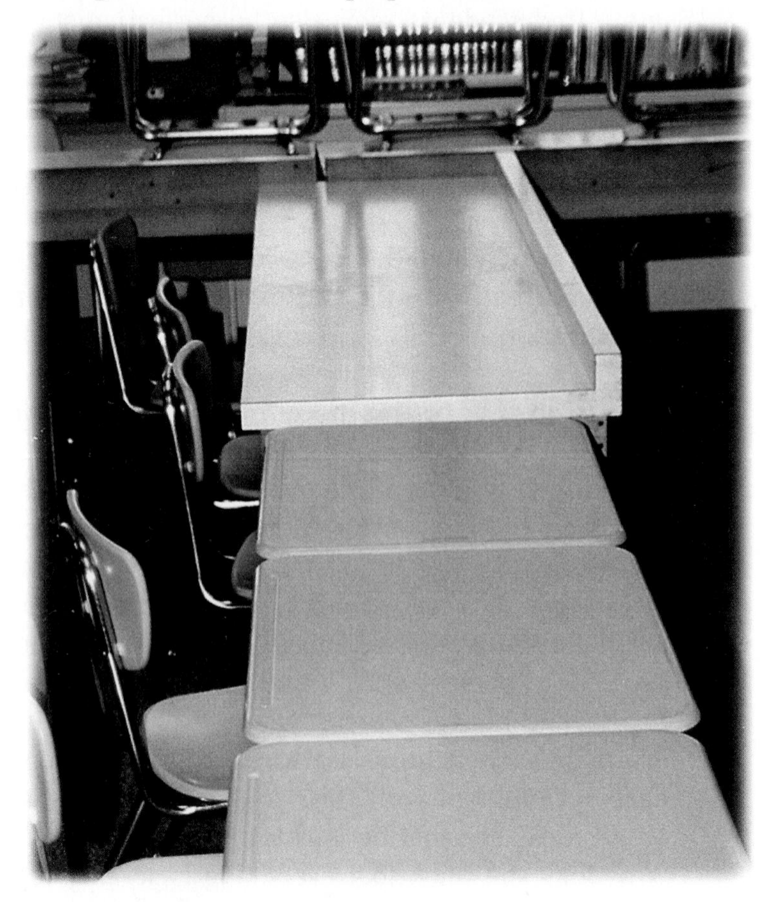

An inexpensive solution is to build tabletops and set them on top of several student desks, creating one large, portable lab table. These tabletops cost less than new furniture, and you can design them to meet specific needs. In my classroom, I have students work in small groups. Therefore, I chose a design that covered three student desks pushed together. Materials are readily available at your local lumberyard. I recommend particleboard and plastic laminate, which is inexpensive and light enough for students to lift.

If you don't have the time or expertise to build tabletops, you may have them built for you. With the approval of your school's maintenance supervisor, a member of the custodial staff may be able to build them for you. Be sure you have a workable design—a table that is large enough for students to work on but small enough to be easily removed and stored.

Securing Lab Equipment

Another concern for teachers is securing the proper equipment for the lab experiments. As every middle school science teacher knows, lab equipment has a way of getting broken or lost or simply disappearing. Replacing all those thermometers, test tubes, and Petri dishes takes time as well as money.

To reduce costs and to get the most out of the equipment you do have, consider that many items can be made using items that would normally be thrown away.

For example, a collection of graduated cylinders is a necessary part of any laboratory. These small containers are expensive and can be easily broken. But, homemade counterparts are easy to make. Using one commercially manufactured set, you can make extras with plastic pill bottles and a permanent marker. Simply fill the graduated cylinder to the desired measurement, transfer the liquid to the pill bottle, and use a marker to carefully label the level of liquid in the pill bottle. Putting your class to work on this project for half an hour will supply students with graduated cylinders for an entire school year. Ask students to bring in empty pill bottles, or ask a local pharmacy to donate them.

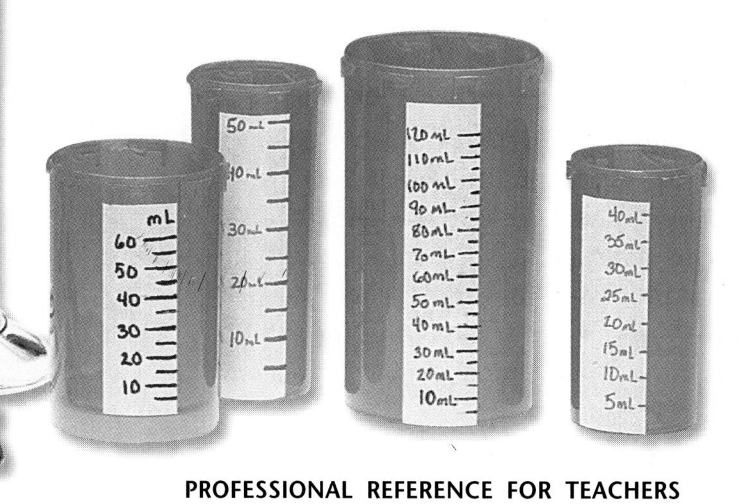

A soft-drink bottle can also be used to make a couple of essential pieces of science lab equipment. The lower half of the bottle can be made into a large measuring container, while the top half can become a "free" funnel, allowing for fewer spills in the lab. Place a coffee filter in the funnel, and you have a filtration device. These funnels are so inexpensive to make that they can be thrown away once they are too soiled.

If You Can't Buy It, Borrow It

Some pieces of lab equipment are just too complex to make and too expensive to buy. The logical solution is to borrow the equipment. Probably the nearest source of such equipment is your own high school science department. Many teachers shy away from this resource because they feel middle school teachers and high school teachers are in competition for school funds and supplies. By pointing out to a high school science teacher that your students are tomorrow's high schoolers and that the students' use of the equipment will make them more prepared for high school science, you may go a long way toward loosening a teacher's grip on some prized equipment.

The willingness of high school teachers to loan materials can be encouraged by the coordination of curricula. This is something many states and school systems are doing anyway. Coordinating the curriculum of the middle school with the curriculum of the high school creates a greater understanding of what content is covered in the respective years, and it promotes cooperation between middle school teachers and high school teachers.

Other good resources for lab equipment and supplies are your local colleges and universities. Many such institutions, especially those with teacher-education programs, are very willing to loan or donate needed equipment. Get to know the professors, and seek their assistance. Professors are usually very willing to loan equipment and donate supplies, and they or their students may come to visit your classroom to talk about their field of expertise or to perform a science demonstration.

Lab equipment can also be shared between neighboring school systems. You might be surprised to learn how much is out there if you are willing to look for it.

One idea is to work together with teachers from adjacent schools to set up a database of loanable lab equipment.

This is similar to an interlibrary-loan system. It takes some effort to get started, and it may be best to begin with an informal arrangement among a small group of teachers. Over time, a list of supplies and equipment available for loan can be compiled. The database can then be made available to a wider group of teachers.

There are negatives to borrowing equipment. It takes time to coordinate and transport equipment back and forth. Also, when borrowing equipment, you must be flexible as to when you will use it. The teachers involved have to discuss the timing of the various units being taught. If cooperating teachers can schedule their classes to stagger instruction, more classrooms can get full use of essential equipment.

Additional Help

There are many additional resources to help teachers create an effective and inexpensive lab program. These resources include television shows, teacher workshops, the Internet, and books.

Science television shows are a readily available source for cheap lab ideas that can be used or adapted for use in the classroom. Especially useful are reruns of *Mr. Wizard* and *Bill Nye the Science Guy.* Both shows contain lab segments geared for children. They are fun and interesting, and they usually require a minimum of supplies and equipment. These labs can be done at home or can easily be adapted and expanded for use in the classroom.

Teacher workshops are also an invaluable source of information on inexpensive labs.

The workshops I have found most useful for developing a lab program in my classroom have been arranged by local schools, colleges, and BOCES (Board of Cooperative Educational Services). These workshops, which are conducted by teachers, emphasize a sharing of ideas and information. Teacher-run programs are an excellent resource for lab ideas because the ideas discussed have been conceived, designed, and tested in the classroom.

During the past few years I have found myself relying more on the Internet for new lab ideas. Many teachers post information about their lab programs on school Web pages. An evening of browsing the Internet, using search engines with key phrases such as "hands-on science" and "middle school science," will produce a huge volume of relevant information.

Of course, the most important resource for new ideas is our own creativity. The more innovative and interesting our lab ideas are, the more our students will benefit. Middle school students, being naturally creative and imaginative themselves, respond enthusiastically to creative labs. This increases students' academic involvement and achievement. Getting your students engaged in meaningful hands-on experiences will also result in fewer discipline problems and lower job-related stress. For me, the most important result of the extra effort I have put into creating an exciting and creative lab program is a job that is still fun after 26 years! ■

The Internet: Realizing the Potential

by David Warlick

David Warlick *is an instructional technology consultant who has been involved in the educational use of computers since 1981. He is a leader in innovative applications of technology, especially the Internet. His work experience combines nearly 10 years as a classroom teacher, 6 years as a central office administrator, and 6 years with the North Carolina State Department of Public Instruction. Warlick develops and conducts numerous curriculum projects over the global network and facilitates education workshops from California to the Netherlands. He is the author of* Raw Materials of the Mind: Teaching and Learning in Information and Technology Rich Schools.

The Dawn of the Internet

In the spring of 1990, only a few weeks after I had acquired access to the Internet, I received an E-mail message announcing a new service called Internet Relay Chat, or IRC. (Today we just call it "chat.") The service was provided by a local university in Raleigh, North Carolina, where I had been living for 2 months. When I returned home that evening, I logged on to the Internet and connected to the service. I entered a chat room and soon began to see messages as I scrolled down the screen—a "live" stream of conversation between four chatters. Because of the nature of their comments and the fact that our university has a respected Computer Science Department, I concluded that my "virtual friends" were local computer science students.

After I joined the conversation, we began to talk about how fifth graders might use "chat" as a part of their access to the Internet. As we continued to share ideas, I could tell that they were getting quite excited about the possibilities. Then one of the chatters suggested that we meet for pizza and continue the discussion in real life (RL). Given my newness to the area, I struggled with the street names and directions they gave me in order to find the pizza parlor. Finally, I asked the students their location. They replied, "Reykjavik, Iceland." I decided to take a rain check on the pizza!

Dissolving Barriers For two weeks I boasted to friends about the night I discussed educational technology over the Internet with people from Iceland for free. Eventually, it occurred to me that this free, worldwide chat arena was not the most important aspect of the experience. For 40 minutes, it did not matter that we were thousands of miles apart from each other, that we spoke different native languages, grew up in different cultures, and would probably never meet.

Thanks to the Internet, we were people sharing ideas and adding to each other's experiences through a technology that hid all but our thoughts, ideas, and knowledge.

When this strange collision of cultures occurred nearly 10 years ago, only a handful of teachers knew of the Internet's existence. No one suspected that the time would come when this type of culture-crossing experience would be a common, everyday experience for many students. Today, many teenagers and younger children consider people from other continents as their friends. They will probably never meet their virtual friends, but their on-line conversations are teaching them how different and yet alike we all are. The opportunity to establish and learn from such worldwide friendships can only better equip our young people to be more globally minded in the world they will one day lead.

Global Network in the Classroom As schools increasingly connect to the global network, teachers are challenged to help students use the technology within the framework of the curriculum to reap academic rewards. The question is what is the best way to use Internet technology in the classroom? The trend has been to continue traditional teaching strategies and to integrate the technology where it obviously fits. For instance, we see fingers tapping a keyboard instead of picking up a pen to write a report, drill-and-practice software replacing flash cards, and the Internet and CD-ROMs replacing bulky encyclopedias and other reference books. Each of these is an example of integrating technology into the classroom, but thus far, it is the technology that has been adapted to the classroom.

Boldly Step Forward! The key to realizing the potential of the Internet and its related tools is to identify the unique qualities of the technology and then take advantage of those qualities by boldly adapting our classroom activities. In this article we will explore three aspects of the Internet that are uniquely suited for helping students learn.

The Internet Is Valuable As An:

1. **Information Resource:** a convenient source of information in a variety of formats and from a variety of perspectives
2. **Arena for Collaboration:** a place that allows interaction with peers, on-line communities, and experts in order to gain knowledge and accomplish goals
3. **Arena for Self-expression:** a place that provides students with real audiences for their work as well as the opportunity to receive authentic feedback from those audiences

1. Information Resource

Perhaps the most obvious value of the Internet to students is that it is a rich and dynamic information resource. In January 1996, there were only 100,000 Web sites on the Internet. Today, we frequently get three times that many Web pages by searching for a single keyword. This indicates a vast and rapidly growing library of information that is readily available to students and teachers.

Finding, evaluating, and selecting the best information is a challenge for even the most skilled Internet user. Unfortunately, this issue cannot be addressed adequately in one article. Links to this week's best search engines would be of little value six months from now. In order to get the most out of the Internet as an information resource, seek out professional development opportunities that will teach you how to search the Internet and examine what you find with a critical eye. Then practice your newly found skills. Mastering these skills yourself lends you the confidence to teach them to your students.

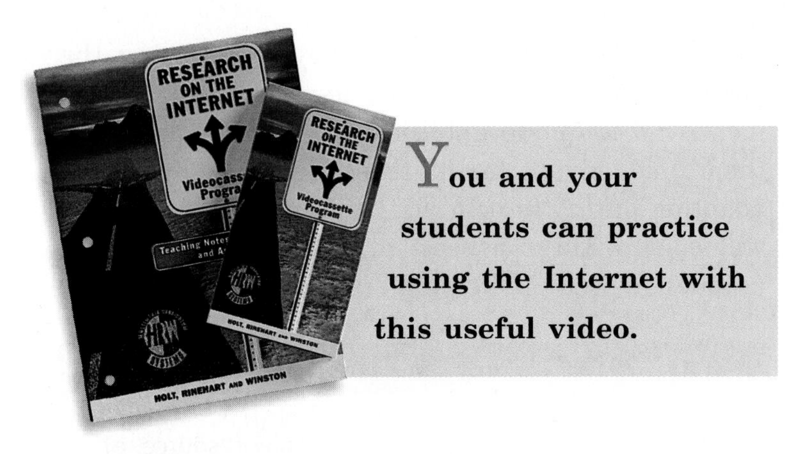

Once you find and retrieve the information you need, what's next?

The greatest advantage that Internet information has over information from traditional print sources is that Internet information is digital.

Internet information comes from a hard disk somewhere on the planet and is copied to your hard drive. From there, you can usually transfer the information to another processing tool, such as word processor, spreadsheet, database, or graphics software, and manipulate the information into the format and presentation that best serves your needs. At right is an example of how you can process information from the Internet for your own use.

Find and Interpret Data in No Time!

You visit a Web site called the Council of the National Seismic System Catalog Search.[1] As the Web page loads, you note the current time on your computer, 10:00 A.M. A form appears, and you fill in the blanks to request data on all earthquakes that occurred during April 1998 that registered greater than 3.0 on the Richter scale. You submit your request, and after a few minutes, a new screen appears that displays the data for over 1,150 earthquakes. You transfer that data into a spreadsheet program, paying particular attention to the location of the earthquakes. Highlighting the latitude and longitude columns, you click the graphing tool and produce a scatter plot, marking the position of each earthquake. Then you paste the scatter plot into a graphics program, where you add text and arrows, labeling the equator, prime meridian, and several geographic regions, and identifying the tectonic plates. After you print this graph that fits your original earthquake specifications, you glance at the time—10:15 A.M. It has taken you a mere 15 minutes to collect digital data from the Internet and produce a teaching tool that helps students see the relationship between earthquakes and the tectonic plates!

[1] http://quake.geo.berkeley.edu/cnss/catalog-search.html

2. Arena for Collaboration

Conversations and group discussions turn the cyberworld into an ongoing global conference on a wide range of subjects. Most people rely on the hypertext, multimedia realm of the World Wide Web and tend to overlook individuals as a viable source of information. Working with individuals over the Internet has its advantages. When someone provides information via E-mail, a chat room, or a discussion forum, the information can be tailored to specific curriculum needs and adapted to the maturity and comprehension level of the students. A person can also elaborate on issues of special interest to students or teachers. We cannot expect this type of special treatment from static Web pages designed for the general public.

Internet collaboration can be approached from three perspectives: (1) collaborating with experts to gain knowledge and insight; (2) collaborating within on-line communities of interest, tapping into their conversations and seeking information from a wide range of experiences and knowledge; and (3) collaborating with peers over the Internet, such as students working with other students through E-mail or other collaborative tools.

Talking to the Experts Let's say that you are preparing a new unit on earthquakes. Your goal is to arrange a way for your students to learn the current issues, concerns, and interests of earthquake scientists. This would provide students with a context for the knowledge that is required to meet your state's education standards. There are a variety of ways to find an earthquake scientist over the Internet, but a consistently valuable place to start is a government Web site. Conveniently, the Web addresses, or URLs (Uniform Resource Locators), for all state government Web sites are almost identical except for the two-letter state abbreviation. For example, if you want to find an expert in California, use the following address: http://www.state.ca.us. Through the California state government site, you can link to education in that state and then to a list of universities. After selecting a specific university, you can locate the Web page for the Department of Geophysics. Here you will find the professors' names, E-mail addresses, areas of expertise, and current research topics. With this information, you can determine which professors might be best suited to answer questions directed toward your curriculum goals and then approach those individuals via E-mail.

Joining an On-line Community On-line communities have many forms. Perhaps the most common and powerful on-line community is the Internet Mailing List, sometimes called a LISTSERV. A LISTSERV is a list of many individual E-mail addresses represented by a more general E-mail address, such as scienceteacher@uncb.edu. An address such as this might connect to a list of over 500 science teachers across the United States. When one message is sent to scienceteacher@uncb.edu, each list member receives a copy of the message. Ongoing conversations take place through these Internet mailing lists on a wide variety of topics, including earthquakes.

You can also search a number of on-line databases to find mailing lists. For instance, you might go to http://www.liszt.com/. Here you would type "earthquake" and then click the search button to identify a number of earthquake-related mailing lists. You would scan the display of lists and learn that the QUAKE-L list is for the discussion of earthquakes and earthquake-related phenomena and that the levels of expertise range from novice to expert. You would also learn how to join the list and participate in the discussion. I joined this list in early 1998 and immediately posted a question about teaching an earthquake unit to middle school students. Answers from earthquake scientists started arriving in less than 30 minutes.

Students Working with Students Asking students to collaborate with each other over the Internet is another valuable technique. By working on a project that requires them to collect information and create some type of visual end-product, students will not only learn the content required by your state's standards but also benefit from an enriching exchange with students from another culture. For example, you might establish a project with a class in Italy, a country that experiences frequent earthquakes. Finding a class in Italy can be quite easy since Web 66—one of the oldest education Web sites on the Internet—has a worldwide registry of schools that have their own Web site. Try typing the following address: http://Web66.umn.edu/. Once the page has loaded, click on the International Registry of Schools, and select a map of Europe. When you select Italy, you should get a list of over 100 schools in that country that have their own Web site. You can then visit some of the sites and send an E-mail message to the Web masters describing your idea and asking that your E-mail be forwarded to the appropriate teacher.

One idea for a project might be to ask the students in Italy to prepare a travel guide explaining to your students how to prepare for earthquakes if they were to visit that Mediterranean country. In turn, your students could prepare a similar document to help Italian students prepare for some unique characteristic of your location.

3. Arena for Self-expression

The relay of information is inseparable from the learning process. In the classroom, teachers relay new information to students, and students learn the information and then demonstrate their new knowledge by relaying it back through reports and tests. But too often there is no real-life context for the students' work because what they produce will be seen only by the teacher and only for a grade. The Internet provides unique opportunities for giving students real audiences for their work.

Research indicates that students write better when publishing to a real audience. In 1986, Moshe Cohen and Margaret Riel conducted a study that sought to measure the difference between writing exclusively for a grade and writing to real and distant audiences via E-mail.[2] Two seventh-grade classes in Israel were selected, and each student was given two writing assignments. Students were informed that their performance on Assignment A would determine their grade for the entire grading period. They were told that Assignment B would not be graded at all but would be e-mailed to a group of fellow seventh graders, in San Diego, California.

When the students finished their work, the papers were coded and shuffled and then given to independent scorers, who did not know which papers were destined for California and which papers were destined for the grade book.

Cohen and Riel found that students wrote better when they were writing to the seventh grade audience half way around the world.

Specifically, they found that the students wrote more, wrote in greater detail, and took greater care with spelling, grammar, and punctuation.

Upon examining the students' work more closely, the researchers discovered that students tend to assume that their teacher already knows about the things they are writing about. Therefore, they are not communicating anything new to the teacher. However, they do not make the same assumptions about their distant audiences. As a result, they understand that they are communicating in an authentic way and take greater care to be more effective in the description of their information.

Bringing It All Together in the Classroom

The three aspects of Internet collaboration described in this article can be used separately or together in the classroom. Students can access information by collaborating with experts and peers and by searching for and evaluating Internet information. Since the information that they collect is digital, students can move the data into other information-processing software, using it as a basis for drawing conclusions or using the computer to assemble new and valuable information products. As students compile their conclusions and assemble the information building blocks that they have collected, they can publish their report through the Internet for people to read.

A wonderful example of the above is the ThinkQuest™ program. This program's Internet site brings together students from around the world to work in teams. These teams develop Web sites that help other students learn something.[3] The value of what students can accomplish on the Internet becomes clear when one realizes that the student-built sites created for ThinkQuest already receive an average of 3,000,000 Web "hits" a day.

---

[2] Cohen, M. and M. Riel, 1989. (Complete references can be found in *Section III: Continuing the Discussion,* page 142.)

[3] Please see the following Web address: http://www.thinkquest.org.

The Internet for Teachers

Right now, providing Internet access to teachers will have a greater educational effect on the classroom than giving every student access to the global network. There are several reasons for this.

First, access to the Internet is still a scarce commodity. While few schools can facilitate large numbers of students on the Internet, many can provide adequate access to teachers.

Second, the use of basic information technology must become a practical working skill for teachers before they can effectively teach it to others. This takes practice. Also, given the speed at which information can be exchanged, teachers can tap into one another's experience and knowledge and support one another in ways that have never before been possible.

Third, the Internet represents a warehouse of teaching materials. Teachers can find and evaluate information, and then bring it into the classroom as handouts, transparencies, video clips to be viewed through the classroom computer, or entire Web sites that can be used without access to the Internet. Teachers are no longer limited to magazines and newspapers for supplemental materials. They have a global library to choose from.

Finally, one of the most important benefits of the Internet for teachers is its ability to facilitate innovation in the classroom. When you log on to the Internet to find resources for your class, you will likely find information that you expected—a Web site about earthquakes, some seismographic printouts, and some pictures. But when effectively researching the Internet in a procedural way, you will also find resources that you probably did not expect. For instance, a procedural search of the Internet might turn up a series of QuickTime virtual-reality files of earthquake damage in Iran. Through these files, students can safely take a virtual stroll down the streets of Annifo, examining the damage and getting a feel for the sheer power of the Earth's shudders. As you find new and unexpected resources, you will naturally think of unique learning experiences for students—learning experiences that would not have occurred to you otherwise.

The Appeal of the Internet

Many of today's students have easy access to cable TV, VCRs, video games, computers, and the Internet. This growing sophistication of our students as savvy media users makes it difficult to attract and hold their attention with chalkboards, activity sheets, and textbooks. But skilled teachers, with vision and access to reliable technology, can compete with the television, video games, and movie theaters by helping students learn and be successful in the future that they will someday inherit.

__Lloyd H. Barrow__ is a professor of science education at the University of Missouri-Columbia, in Columbia, Missouri. Some of the areas Barrow is interested in are staff development in science, Earth science education, and preservice elementary science teacher preparation. He has presented at regional and national NSTA conventions for more than 20 years, and he served as the research director for NSTA from 1996–1998. In addition, Barrow has 9 years experience as a science teacher for grades 7–9 and 3 years of experience as a science coordinator for grades K–12. His experience with family science night programs spans more than 6 years.

Organizing a Family Science Night at Your School

by Lloyd H. Barrow, Ph.D.

Picture an event at your school in which parents and their children participate in hands-on activities to test ideas, solve problems, and learn about science. Imagine a forum in which assumptions are set aside and discovery is the rule and parents and children surprise one another with their determination and teamwork. Sound impossible? Well, it's not. What follows explains how to organize a successful family science night at your school.

What Are Family Science Nights?

Family science nights are catching on in schools all over the country as a way of bringing families together to learn about science. Unlike science fairs, in which students share their expertise with judges and visitors, everyone involved in a family science night plays a role in scientific discovery. While parents and students explore the world of science by participating in hands-on activities, teachers perform demonstrations and answer questions. The end results of such an event are your students' increased confidence in their ability to do science and more parental involvement in science education. In addition, everyone usually has a lot of fun!

So how do you organize a successful family night at your school? It is easier than it sounds, especially if you follow these simple steps:

- ★ **Find a colleague**
- ★ **Choose a theme**
- ★ **Hold a meeting**
- ★ **Choose activities**
- ★ **Plan the program**
- ★ **Invite the families**
- ★ **Request materials**
- ★ **Encourage participation**
- ★ **Keep up the interest**
- ★ **Enjoy the big night**

Find a Colleague and Choose a Theme

The first step in organizing a family science night is to find a fellow teacher who is willing to help you with planning. Make certain that this teacher is not only willing but committed to helping you. Together you will choose a theme for the family science night. The goal is to select a theme that is broad enough to include a variety of activities from the major branches of science—life science, Earth science, and physical science. Keep in mind that every activity you choose must relate to the theme, so refrain from making the theme too narrow. Some possible themes are "Science in Our Environment," "How Things Work," "Stars and Planets," and "Water in Science." Once you've chosen a theme, it is important to meet with the building principal to establish his or her support for the event.

Hold a Meeting and Choose Activities

The next step is to invite other teachers to attend a planning meeting for the family science night. At this meeting a number of other teachers will likely volunteer to join your efforts and see them to fruition.

During this meeting you should pick a date for the event, or at least narrow it down to a few possible dates. You will need the building principal's approval for the date as well, so presenting a few possibilities is a good idea. Don't spend too much time picking dates; it is best to keep this meeting short and use the bulk of the time for coming up with activity ideas. Because the activities you choose are the focal point of a family science night, the more time you can invest in coming up with ideas the better.

Conduct a brainstorming session during the meeting in order to generate a list of activity ideas. Then enlist the help of one or two group members who can follow up on the ideas and further develop the list. There are many resources in the library and on the Internet that you can use to find a variety of activities.

In order to pick the best activities among the many you are likely to encounter, consider the following questions:

- **Does the activity focus on problem solving?**
- **Will the activity appeal to boys and girls and adults?**
- **Is the activity a good springboard for families to do further exploration at home?**
- **Will the activity be able to accommodate teams of two or three family members?**

When you find the ideal activity, you will be able to answer the above questions with a resounding yes. Also, it is best to keep the activities simple in order to keep costs down and minimize cleanup time. Choose one or two activities that you can present to the attendees as a warm-up for the evening's events. These activities might be impressive demonstrations that take longer to set up and use more expensive materials. Then choose six additional activities for families to perform on their own. These activities should require about 15 minutes each to complete.

Parents will enjoy the opportunity to watch their children discover new and exciting things about science.

Given the variety of ages and backgrounds of the participants in a family science night, it is important that all the activities are gender-neutral and assume little or no prior understanding of the subject matter. However, even the most knowledgeable parents can come away with valuable experiences from a family science night. Parents who don't learn new scientific information will enjoy the opportunity to watch their children discover new and exciting things about science.

Plan the Program

When it comes to planning the program, consider opening the event with a presentation, such as a brief slide show or video; a large group activity such as a physics circus (a performance that uses physics concepts to pique interest); or one of the warm-up activities you chose for this very purpose. Even if there is no opening presentation, plan to have a meeting at the beginning of the evening so that the rules can be explained to all participants.

The structure of your program will vary depending upon the room or rooms you choose to use. As a result, there are numerous ways to plan the program. A simple plan for structuring the evening is to use a single room, such as your school's gymnasium or cafeteria. At the start of the evening families are separated into two large groups by last name. These groups will meet at opposite sides of the gym to allow you and a fellow teacher to present the warm-up activities to a more manageable number of people. During the introduction and demonstration, try to focus on the drama and wonder inherent in science as it relates to the activities. The goal is to leave the families eager to perform the activities. Once the introduction and demonstrations are complete, families will move to activity stations set up around the gym that are equipped with all the materials needed to perform the activity. At this point families are free to move at their own pace through the activities, and you and your fellow teachers can circulate through the gym and answer any questions that might arise. As an alternative, you may choose to set up the activities in individual classrooms and have families circulate from room to room.

It is a good idea to keep your first family science night as simple as possible but it is important to make sure it is well adapted to your surroundings. Once you have a definite plan in place for the evening, you are now ready to write a formal invitation to parents and request materials.

Invite the Families and Request Materials

The next step is to design a letter that will serve the dual purpose of inviting families to the family science night and requesting donations for materials. Send out these letters well in advance—about 4 or 5 weeks before the event—and include a tear-off portion so that parents can respond. The number of responses will help you determine the quantity of materials you will need. It will also give you enough time to assess whether you have enough materials and make arrangements to procure more if you are running short.

The first half of the letter you send home with students should be the invitation to the event. Probably the most important point to make on the invitation, after the date and time, is the fact that all students must be accompanied by a parent or guardian. There should also be a space for designating just how many people will be attending.

Ideally, students will bring their parents or guardians, but students should be encouraged to attend with another adult if their parents or guardians are unable to attend. Adult family members, siblings of a responsible age, or neighbors can also participate. As long as students are interacting one-on-one with an adult, they will be able to participate fully in the event.

Since you have already planned the program, you should be able to draw up a complete list of materials you will need. This list should include every item you will need to perform the activities, but don't be concerned with exact amounts at the moment. Defining your needs will allow you to be specific when asking for donations of these materials.

The letter might read something like this:

The honor of your presence is requested on

Tuesday, March 16
6:00 to 8:00 P.M.
at **Steele Middle School**

for a

Family Science Night

about

Water in Science!

Please note: All students must be accompanied by a parent or guardian.

My family will/will not (please circle one) be attending the Family Science Night on March 16.

The total number of people attending is ______ .
(please fill in the appropriate number)

(please cut along the dotted line and return by February 19)

Dear Family,

We hope you can join us the evening of March 16 for our very first Family Science Night. The theme of the event is *Water in Science!* and should prove to be some good, clean (if not a little wet!) fun exploring the world of science. You will have the opportunity to perform your very own experiments in a comfortable and relaxed atmosphere.

In order to hold such an event we need some help from you. What follows is a list of materials that we need. If you have any items to donate, please bring them to the school office before March 9.

Parents of Steele Middle School students have always been eager to support our special events. The faculty would like to thank you for all of your support and any donations you might be able to make towards our first Family Science Night.

Materials List: disposable plastic spoons; disposable plastic cups; 1 and 2 L plastic soda bottles; coffee filters; drinking straws; pipe cleaners; masking tape; plastic containers from cream cheese, margarine, and cottage cheese; baking soda; baking powder; food coloring; colored balloons

Thank You!

Encourage Event Participation

How do we involve busy parents who do not have time to accompany their children to school, sports club, or youth group events? One idea is to use a variety of incentives, such as food or contests with prizes. These types of incentives can also help to create a more festive atmosphere during the event. Possible sources for these incentives are your students' families and local businesses.

Some of your students' parents may own a business or hold a position in a company that would be willing to donate prizes or supplies for the activities.

These businesses may be willing to donate in exchange for an advertisement in the event program or permission to set up a booth during the event. Another source of funding are small grants that are often made available by community organizations and local businesses. You may find it worthwhile to take time to make some telephone calls with these possibilities in mind.[1]

Keep Up the Interest

Once you get parents to attend a family science night, there is still the potential for them to socialize while the students engage in the activities. One way to keep parents actively involved is to provide them with a handout that briefly explains the science behind each activity, so they can answer questions from their children. On the handout you can also list related scientific information, challenging questions, and any other items that might spark conversation between parent and student during the course of the evening. Don't hesitate to include a list of resources on the handout that families can use in order to get more information about the theme of your science night. Such resources might encourage families to pursue further study or even perform activities at home.

Another way to engage parents is to entrust them with the more complicated parts of the activity or the parts that require delicate motor skills. These directions should be added to the parents' handout and posted at each activity station.

[1] For information about seeking larger grants, please see the article "Applying for Education Grants," by Ernest W. Brewer and Connie Hollingsworth, on page 124.

Enjoy the Big Night

Hours before the big night you should prepare the room(s) you will be using by setting up the demonstrations and the activity stations. Be certain you have enough supplies based on the returned invitations, but always plan for a few "walk-in" families. Start with at least five complete setups per activity station. By taking the time to mix and count the necessary items beforehand, you will ensure a more enjoyable time for the participants. This allows the participants to focus on performing the activities and enjoying themselves, rather than on setup.

After the initial demonstrations, you and the other volunteers should circulate around the room, in case there are any questions. You may need to restock an activity station's supplies or help a family set up an activity correctly. For the most part, however, this is the time to enjoy the fun as you watch students and parents delve into the world of science. It is also a chance to meet family members you may not have had the opportunity to meet before.

> **This event might be the bridge that brings students, teachers, and parents together in a way you never thought possible.**

As the evening draws to a close and you begin to say good-bye to families, ask them what they liked about the event and what changes or improvements they might suggest. Ask those parents who really enjoyed the event to consider helping in the production of future science nights.

Building Bridges

A family science night takes careful planning, but you need not do it alone. Tap into the ideas and efforts of fellow teachers and your students' parents. Remember that your first family science night need not be a fancy and complicated production. Starting out small leaves plenty of room for improvement for the next event, so consider each one a learning experience. The goal of the evening is to get your students and their parents actively involved in science in a comfortable environment. Such an event just might be the bridge that brings students, teachers, and parents together in a way you never thought possible. ■

Teaching and Learning Science Through

by Carol M. Santa, Ph.D., and Lynn T. Havens

Carol M. Santa *has been an elementary school teacher, reading specialist, Language Arts Coordinator, university professor, and author. She is currently the co-owner and Education Director of Montana Academy, a residential school for troubled teenagers.* **Lynn T. Havens** *has taught math and science at both the middle and high school levels. She has authored several articles and journal chapters on the application of strategies in math and science education. Together, Santa and Havens developed Project CRISS (CReating Independence through Student-owned Strategies), a learning-strategies program that focuses on reading, writing, and studying across the curriculum. Project CRISS is being used in schools across the United States and Europe.*

How can teachers help students develop the skills to observe and think scientifically? Writing is one of the most powerful, effective skills a young science student can develop. Writing about science not only helps students organize their thoughts and questions; it allows them to take an active role in the learning process as well.[1] It penetrates the external shell of memorized facts and superficial understanding and helps young minds tap into the core of learning.

In this article we will discuss three of the most successful types of writing assignments: explaining scientific phenomena, learning logs, and scientific reports.

[1] Santa, C.M., and L.T. Havens, 1991. (Complete references can be found in *Section III: Continuing the Discussion,* page 143.)

Benefits of Writing in the Science Classroom

***Writing* incorporates previous knowledge.**
Integrating new information with background knowledge is fundamental to science learning, and writing can help with this process. Writing about a topic *before* reading the lesson summons prior knowledge, which is then easily incorporated with new information.

***Writing* helps students become metacognitive.**
Good readers monitor their comprehension. They know when they understand and what to do when they don't. Writing helps students gain this awareness, in part by providing a means of measuring their own knowledge. Students cannot write clearly about something if they do not understand it.

***Writing* encourages active involvement in learning.**
Effective learning is not something we can do for our students; it requires initiative. Too often students remain passive, like empty vessels waiting for teachers to fill them with knowledge. It is impossible to remain passive as a writer, however. When writing about observations or a reading assignment, students are drawn into the learning process as participants.

***Writing* builds organization skills.**
Writing helps students see clusters of information and hierarchies of ideas. As students build systems of organization, they make new information their own.

Explaining Scientific Phenomena

Asking students to assume the role of teacher by providing their own oral and written explanations for phenomena is another way to encourage active learning. We all remember our first year of teaching when we finally understood our content because we had to teach it. In doing so, we became active learners. As most of us have discovered, it is impossible to teach effectively while being a passive learner. The rule works in reverse as well. When students work in pairs, explain concepts to each other, and then put their explanations on paper, they learn actively. This strategy is an adaptation of ideas drawn from cooperative learning[2] and reciprocal teaching,[3] both of which have proved highly successful.

To encourage clear and complete explanations, ask students to write for someone who knows nothing about the content. When students write for you, their explanations lack precision and clarity because they know you already understand the content. Why should they write detailed explanations for a knowing audience? If the intended audience knows nothing, however, explanations must be more thorough and clear. Allow students to choose their own audience, such as a parent, a younger sibling, or a friend.

You can also encourage a variety of roles and formats for writing exercises. Instead of always writing as themselves, students can take on other roles, such as a blood cell, a pollen grain, or an amino acid. You may also vary the writing format from an essay to a letter, editorial, diary, obituary, or memo.

When students are allowed to choose their audience and format, their writing becomes livelier and more fun to read.

Writing experts tell us that students need to choose their own topics for writing.[4] School writing should mirror the work of real writers, covering topics of personal value for real audiences. Yet, as science teachers, we want our students to write about important science content. Allowing students to choose their own roles, formats, and audiences when explaining a topic provides some freedom of choice even when the topic is constrained. Explanations become livelier and more fun to read. And students learn much more in the process.

---

[2] Danscreau, D.F., 1985.

[3] Palinesar, A.S., and A.L. Brown, 1986.

[4] Calkins, L., 1987; also see Graves, D., 1983.

Learning Logs

The learning log is a student's written journal of questions and thoughts about the material covered in class. Entries may be questions about vocabulary, difficult concepts, or lab results. It should be a somewhat informal, even messy, record of their thoughts. The important thing is that students commit these thoughts to writing, which encourages them to remember, puzzle through, and perhaps find the answers to their questions. If you keep a personal log (which we also recommend), share it with the class. This can add a real sense of relevance to the practice.

To get your students started with their log, try using a reading assignment as a springboard. For instance, when introducing a new chapter in the textbook, have your students examine the title, subheadings, and chapter questions before they read each section of the text. Have them brainstorm what they already know (or think they know) about the topic and jot it down in their log. After they've read the text, ask them to write a second log entry on what they learned. Many students will be surprised to see their misconceptions refuted in their own handwriting!

Take a look at the example shown on the next page—the reading entries of one student, Misha, from a section on flower reproduction. Misha's teacher concluded the lesson by asking her students how using their log helped them learn. If students understand why writing is effective, they are more likely to embrace it as a learning strategy. By providing opportunities for writing and talking about why it works, you can often convince students to write on their own.

Recording Observations

In addition to using learning logs for process discussions and as part of pre- and post-reading activities, students can also use them to record scientific observations. Observation is a cornerstone of science. Writing provides the feedback and direction students need to sharpen their observation skills. We want our students to find nuances in nature, to hypothesize and explain their observations. When students write about what they see, they are forced to organize their observations, often allowing them to see even more. Moreover, as students write their observations, they discover meaning in what they see.

See how Nina recorded her observations of a flower on the next page. Notice how easily Nina's observations led to questions. Her learning log set the stage for motivated learning. She wanted to know what her observations meant and then generated her own hypotheses, opening her mind to further exploration.

Documenting Learning

While learning logs are very useful as observation tools, they can also be used in other ways. For example, after your students watch a film, complete an experiment, or listen to an oral presentation, allow some time for writing. Ask, "What struck you as important in this session? What do you want to remember?" You might want to participate in these assignments, writing in your log as your students write in theirs. Then you can read your entries aloud and invite students to do the same.

Logs provide opportunities to write informally and to explore content. The idea is to write without fear of teacher interrogation or a hemorrhaging red pen. The writer is always the primary audience. When students used logs in our classrooms, we never graded them, although we usually gave points for completion. We do recommend that you read the logs, because it will give you a better understanding of your students' knowledge. It can also provide important data about your presentation—their questions and confusions can guide your instructional planning.

One colleague asks his students to leave space for him to write back. Student entries trigger his own ideas, and he can't resist responding. Sometimes his responses are questions, sometimes clarifying comments about their observations. More often he nudges with ideas for further exploration. This personal dialogue provides a strong feeling of individual attention for the student.

How do you nurture the development of students' learning logs? First, students must feel safe for their logs to be effective. Entries must be spared from red-pen intimidation and criticism. Second, students must write frequently. Set aside time in class for writing. Keep your own log and model your own writing after your students'. If students see that you take writing seriously, they will begin to believe in their own need to write. Third, write back to students in their logs. As you converse with students through writing, their responses will become more exploratory. Your encouraging comments make log writing feel safe.

Pre-reading Entry

In this chapter I am going to learn about flower reproduction. I know that flowers have male and female parts. I think that these parts are inside the flower. To see them you have to pull aside the petals. I think petals probably protect the reproductive parts, but I am not sure. I remember something about separate flowers for males and females, but I think many flowers have both parts on the same flower. I'm pretty sure you need to have at least two plants before they can reproduce.

Observation Entry

Right below the outside of the flower the stem thickens into a little case or holder for supporting the flower. It looks almost like a crown. From this crown-like form is a ring of petals. The petals are in two layers. The lower petals are greener and look more like the stem than the upper petals. I wonder what use all of these petals have? I know they might attract bees for pollination. It could be that their function is more for protection of the fragile internal parts of the flowers.

Next, I see a ring of tiny petal-like parts in the middle of the flower. There are four of these, and each tiny petal has two sharp spires on each side. I wonder if these are the stamen and pistils? They look very different from the surrounding petals—far more delicate.

Post-reading Entry

I learned that stamens are the male parts of the flower. The stamen produces the pollen. The female part is the pistil. At the bottom of the pistil is the ovary. Plants have eggs just like humans. The eggs are kept in the ovary. I still am not sure how pollen gets to the female part. Do bees do all this work, or are there other ways to pollinate? I was right, sometimes male and female parts are on separate flowers. These are called incomplete flowers. Complete flowers have both male and female parts on the same flower. I also learned that with complete flowers just one plant can reproduce itself. So I was partially wrong thinking it always took two plants to reproduce.

Conclusion Entry

I felt more interested in reading about flowers because I thought about it before I read. I was surprised I knew as much about flowers before reading. Writing helped me realize what I already knew. It made me more curious about reading because I wanted to know if I was right. Knowing that I was going to have to write when I finished reading made me read more carefully. I got more out of my reading by writing. I wasn't very happy about doing it at first, but it helped.

Laboratory Report Guidelines

1. **Purpose.** Why are you doing this lab?
2. **Problem.** What problem are you investigating?
3. **Hypothesis.** What do you think the outcome of this lab will be? Think about what you already know about the topic, and make an educated guess about the outcome.
4. **Materials.** What do you need to do this lab?
5. **Procedure.** How will you do this lab?
6. **Data or Results.** Draw, record, and chart all detailed observations.
7. **Analysis or Conclusion.** Reread your problem statement. Did your results resolve the problem? Explain why, using data from the experiment. If the results did not resolve the problem, explain why.
8. **Class Conclusion.** Final conclusion after class discussion. Modify your earlier conclusion if necessary.

Scientific Reports

We want our students to think like scientists, to be able to define and analyze problems and formulate conclusions. Having students conduct and write about experiments is an effective way to learn and practice these skills. When students write scientific reports they must organize their thoughts and consider how the various parts of an experiment work together. Writing helps them synthesize their ideas.

We begin to teach the scientific method by having the class solve simple problems together, following written laboratory guidelines. Responses to the guidelines become part of the laboratory report.

Start with what your class already knows about a topic. Have them generate one or two questions to investigate. Students will recall knowledge related to the topic and write hypotheses as explanations. Then ask students to brainstorm ways to test the hypotheses, develop an experiment, and list procedures. Once they complete the experiment and record the results, have them answer their original questions. Did their results answer the questions? If not, ask them to come up with alternate ways to examine the issues. Finally, have the students draft a laboratory report from their notes. Stress that these need to be written without clutter, so that another scientist could replicate their results without confusion.

A colleague of ours has her classes do several reports in groups until they feel comfortable enough with the process to write their own. She then distributes copies of the final reports to the class for evaluation. From this evaluation the class develops a checklist of guidelines for writing and grading subsequent reports. This process of development is perhaps even more beneficial than the final product

Laboratory Report Checklist

1. Purpose

- ✔ Have I explained why I am doing this experiment?
- ✔ Did I conclude this section with a hypothesis?

2. Materials and Procedure

- ✔ Did I explain the materials and procedure?
- ✔ Did I explain the steps used to test my hypothesis?

3. Results

- ✔ Did I present the data from the experiment?
- ✔ Are the differences among variables clearly presented?

4. Conclusions

- ✔ Are my conclusions directly based on the data?
- ✔ Did I refer to the hypothesis?
- ✔ If the data did not support my hypothesis, did I provide some reasons for the discrepancy?

because it gives students ownership of their evaluation instrument. After writing reports in a laboratory group and evaluating the reports as a class, most students are ready to begin writing on their own. Before turning in their reports, students read each other's drafts in reaction groups. One student reads his or her work to the other two students, who listen and provide initial reactions based on the checklist. The students then revise their drafts and hand them in. You may find it helpful to take on the role of a student and demonstrate the reaction group process for the class. We also recommend that you develop reaction group rules to ensure that these sessions are productive.

In assigning scientific reports, remember that the goal is to teach scientific process as well as writing skills. If the reports are unclear, the students probably haven't thought through the procedure. If a student cannot write a logical hypothesis and then support it or refute it, he has missed the main point of the experiment. Clear writing is a sign that you and your students are working together successfully.

Conclusions

There was a time when many science teachers thought writing had no place in their classrooms. We had enough trouble just covering the content. We certainly did not welcome the extra duty of teaching writing.

Our science classrooms are far richer now. Learning logs have added vitality to learning. Writing has also empowered students to grasp the complexities of the scientific method. Their thinking travels from the problem to the conclusion and, as they write, they internalize scientific patterns of thought. Finally, we can be sure our students understand concepts when they can teach them to their peers.

As we examine our growth as teachers, we see that our knowledge is similar to the bud of a flower. At the beginning of our teaching careers we are still emerging. Fortunately, we have our classrooms where we can watch our students read, write, talk, and experiment. By continuing to take our nutrients from them, we will surely grow and develop into magnificent flowers. ■

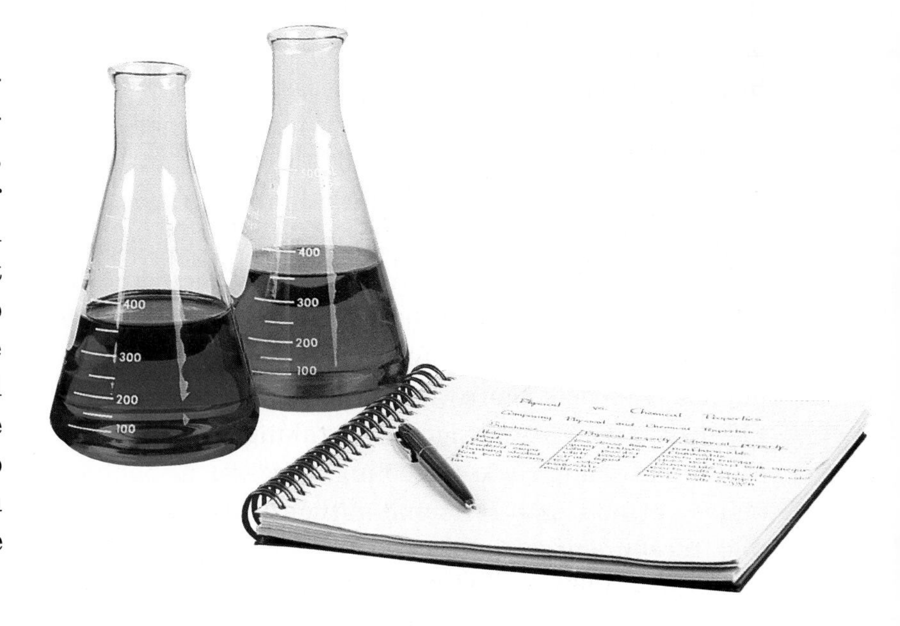

Assessment That Emphasizes LEARNING

by Sandra L. Schurr, Ph.D.

Sandra Schurr *is the Director of the National Resource Center for Middle Grades/High School Education at the University of South Florida, in Tampa, Florida, as well as a faculty member in the Secondary Education Department. Among her many publications are* The Definitive Middle School Guide, ABC's of Classroom Evaluation, *and* Teaching at the Middle Level: A Professional Handbook. *Schurr is also a consultant for assessment, program evaluation, alternative teaching, and middle level/high school restructuring for federal and state education programs.*

What Is Authentic Assessment?

The trend today in middle school assessment is to move toward more realistic types of measurement that focus less on the recall of information and more on the processing of information. Authentic assessment, or performance-based assessment, allows students to demonstrate skills they have learned in a realistic manner. It considers not only the knowledge and skills students have acquired but how they have acquired them. Authentic assessment evaluates students' ability to communicate and apply their knowledge and skills. This method helps students translate the information they learn to the challenges they meet outside of school more than traditional examination methods. Though traditional test-taking measures are still essential to the schooling process, a balance between traditional tests and authentic assessment is now recommended.

Authentic assessment is a type of student evaluation that attempts to make the testing process more realistic and relevant. There are three major forms of authentic assessment—product, performance, and portfolio—and they are defined as follows:

1. **Product assessment** evaluates whether a skill has been applied or some concept has been learned. The actual "products" can range from videotapes, audio tapes, and exhibits to scripts, manuals, and reports.

2. **Performance assessment** is based more on the process the student uses than on the final product or outcome. It relies on the professional judgment of assessors who observe the student performing a predetermined task. Performances can range from oral reports or speeches to scientific demonstrations and poetry readings.

3. **Portfolio assessment** is based on a meaningful collection of student work that exhibits the student's overall efforts, progress, and achievements in one or more subject areas. Portfolio contents can range from paper-and-pencil tests and worksheets to creative writing pieces, drawings, or graphs.

Benefits of Authentic Assessment

Benefits of product, performance, and portfolio assessment include the following:

- collaboration between student and teacher
- features student's individual research and knowledge
- acknowledges different learning styles and interests
- avoids unfair comparisons
- audience extends beyond the teacher
- student has foreknowledge of the type of questions and tasks involved in the assessment
- incorporates a multifaceted scoring system
- incorporates self-assessment

Product and Performance Assessment

Of the three forms of authentic assessment, product and performance are the most closely linked in terms of the benefits they provide students. Also, most of what is created for a product or performance assessment can become part of a portfolio assessment, perhaps for an end-of-semester evaluation in a particular subject area.

Product and performance assessments are important tools because they:

- allow students to show originality and creativity that go beyond what is taught.
- incorporate all or most levels of Bloom's Taxonomy of Cognitive Development in a single outcome.
- reflect growth in social and academic skills and attitudes that are not easily reflected in paper-and-pencil tests.
- motivate and engage students who are reluctant learners and performers in school.
- make learning more relevant and memorable for students.
- demonstrate students' knowledge and abilities in a concrete way.
- allow for easier integration of reading, writing, and speaking skills.
- give students more time and flexibility to do more thoughtful work.
- permit students to interact collaboratively with other students.
- may be adapted to students with varied learning styles.

Authentic assessment evaluates students' ability to communicate and apply their knowledge and skills.

Portfolio Assessment

Portfolios are important assessment tools because they:

- are tools for discussion.
- provide opportunities for students to demonstrate what they know and what they can do.
- allow students to reflect on their work.
- document the growth of a student's learning over time.
- cater to alternative student learning styles and multiple intelligences.
- allow students to make decisions about what to include or exclude.
- make it easier for students to make connections and transfers between prior knowledge and new learning.

Portfolio Tips

If you decide to use portfolio assessment, answering questions such as the following before you begin will help you get the best results:

- What are the major purposes of a portfolio for both student and teacher?
- How should the portfolio pieces be selected?
- Should the number of work samples be limited?
- What specific pieces should be included in the portfolio?
- How do the pieces represent the student's acquisition of content and skills?
- How should the contents of the portfolio be evaluated?
- How should the work samples be organized?
- How can the portfolios be shared through conferences between student, teacher, and parent?

Evaluating Authentic Assessment

The best way to evaluate a product, performance, or portfolio assessment is by using a rubric. Thus the development of a rubric is an important element of authentic assessment. Rubrics are a series of narrative statements that describe the expected levels of quality for a product, performance, or portfolio assessment. A rubric can be a list of narrative statements or a matrix of narrative statements. The rubric can also be developed to analyze a list of specific criteria for a small part of a project or for a complete and final project. The most effective rubrics are those created by teachers and students as a collaborative effort.

Rubric for Reports and Presentations

Possible Points	Scientific Thought (40 points possible)
40-36	Complete understanding of topic; topic extensively researched; variety of primary and secondary sources used and cited; proper and effective use of scientific vocabulary and terminology
35-31	Good understanding of topic; topic well researched; a variety of sources used and cited; good use of scientific vocabulary and terminology
30-26	Acceptable understanding of topic; adequate research evident; sources cited; adequate use of scientific terms
25-21	Poor understanding of topic; inadequate research; little use of scientific terms
20-10	Lacks an understanding of topic; very little research, if any; incorrect use of scientific terms
Possible Points	**Oral Presentation (30 points possible)**
30-27	Clear, concise, engaging presentation, well supported by use of multisensory aids; scientific content effectively communicated to peer group
26-23	Well-organized, interesting, confident presentation supported by multisensory aids; scientific content communicated to peer group
22-19	Presentation acceptable; only modestly effective in communicating science content to peer group
18-16	Presentation lacks clarity and organization; ineffective in communicating science content to peer group
15-5	Poor presentation; does not communicate science content to peer group
Possible Points	**Exhibit or Display (30 points possible)**
30-27	Exhibit layout self-explanatory, and successfully incorporates a multisensory approach; creative use of materials
26-23	Layout logical, concise, and can be followed easily; materials used in exhibit appropriate and effective
22-19	Acceptable layout of exhibit; materials used appropriately
18-16	Organization of layout could be improved; better materials could have been chosen
15-5	Exhibit layout lacks organization and is difficult to understand; poor and ineffective use of materials

Benefits of Collaborative Rubrics

Self-evaluation techniques can:

- place the burden of assessment on the individual.
- answer the student's two most basic questions: "How am I doing?" and "Where do I go from here?"
- provide the basis for agreement between student and teacher on academic priorities.
- improve effectiveness, as opposed to efficiency, in the schooling process.
- encourage objective analysis of one's own attitudes and aptitudes.
- relate progress to performance by answering such questions as "Are we doing the right things?" and "Are we doing the right things 'right'?"
- encourage individual goal setting.
- promote a feeling of personal growth, responsibility, and accomplishment.
- acknowledge differences in learning styles.

Testing Formats that Challenge Students!

Although today's trend in assessment is toward the processing of information, which uses authentic assessment, traditional paper-and-pencil tests are still important elements of any effective classroom-assessment program. However, teachers should use a variety of testing formats whenever possible. While multiple choice, short answer, true/false, and matching test questions are appropriate in many subject areas, alternatives should be considered in designing a varied and comprehensive traditional testing format. This is especially important when creating tests that measure higher-order thinking skills, since traditional types of test questions tend to focus on recall and comprehension. The four alternative testing formats outlined on the pages that follow are popular with students because the test structures allow for student choice, input, creativity, and student-generated responses. The four testing formats include a Bloom test, a fact-finding test, a hands-on test, and a cause-and-effect test. Read on to learn more!

A Bloom Test

This type of alternative assessment involves creating a test made up of six questions or tasks—one for each level of Bloom's Taxonomy.[1] Assign 10 points each for knowledge and comprehension tasks. Assign 15 points each for the application and analysis level tasks. Assign 25 points each for synthesis and evaluation tasks. The result is a test with six questions totaling 100 points. See the brief review of Bloom's Taxonomy below.

Review of Bloom's Taxonomy:

KNOWLEDGE: Students learn terms, facts, methods and procedures, and concepts.

COMPREHENSION: Students understand uses and implications of terms, methods, and concepts.

APPLICATION: Students practice theory, solve problems, and use information in new situations.

ANALYSIS: Students analyze structure, recognize assumptions and poor logic, and evaluate relevancy.

SYNTHESIS: Students write themes, present speeches, plan experiments, and put information together in a new and creative way.

EVALUATION: Students set standards, judge with purpose, and accept or reject on the basis of criteria.

Here is a variation of a Bloom test: Develop a bank of five or six questions at each level of Bloom's Taxonomy using the same point-value system outlined above. Allow students to select their own questions, instructing them to choose items whose total value is 100.

[1] Bloom, B., 1950. (Complete references can be found in *Section III: Continuing the Discussion,* page 143.)

During a fact-finding test, each student both teaches and learns new information.

A Fact-Finding Test

Prepare a set of information cards on a topic to be studied in class, making certain you have a different card for each student. The information cards should contain a relevant paragraph or two on an important concept or issue associated with the topic. You may want to create duplicate cards for concepts that are especially important or difficult. Each student receives an information card to read and study for 5–10 minutes.

Once the students understand their assigned concept, they stand and circulate, pausing just long enough to share their information informally with one another. During this interaction, each student both teaches and learns new information. Students may use the cards only as a reference; they may not read directly from the card. After approximately 30 minutes, students sit and respond in writing with five summary statements. Sample summary statements include the following:

1. The information from my card I taught to others was . . .
2. Three new things I learned about the topic from other students are . . .
3. One part of the topic that I would like to know more about is . . .

Collect student summaries for review and grading.

A Hands-on Test

For this test, identify nine key terms or concepts related to a specific topic and write them on the chalkboard. Students then take nine index cards and write one term or concept on each card. Next they randomly place the completed cards on their desk in a 3×3 grid. Students then write a content-based statement for each horizontal row (three statements), for each vertical row (three statements), and for each diagonal row (two statements). The statements should show how the three terms or concepts are related. The terms in the statement may appear in any order and may be used only once; students may add words in order to convey the interrelationship of the ideas with one another. When students are finished or time runs out, have them submit their statements to you for review and grading.

A Cause-and-Effect Test

This testing format requires the student to explore a variety of causes and effects in situations that are somehow related and are part of an overall theme or subject. For example, here are a few cause-and-effect questions that might be asked on the topics of force and motion:

- List two reasons why you might want to measure force.
- List two reasons why you might want to measure motion.
- List two reasons why a ball might fall through the air.

After students have had a chance to demonstrate their understanding of causes and effects, collect their answers for review and grading.

Test Preparation Tip

Relieving "test anxiety" is essential to student success. Students will likely have questions such as the following. To help put them at ease, answer their questions prior to administering the assessment.

1. Will this be a quiz, test, or exam?
2. What type of assessment will it be (take-home, oral, collaborative, individual, or open-book)?
3. What information will be covered (notes, textbook, lecture, outside readings, discussions)?
4. What kind of questions will be asked (multiple choice, essay, fill-in-the-blank, true-false, or short answer)?
5. How much time will I be given to complete the assessment?
6. Do the answers have to be written in complete sentences, and will spelling count?
7. How can I best prepare for the assessment?
8. How will the assessment be graded?

Armed with the above information, students can focus their energies on learning the required material rather than on worrying about what type of testing format they will encounter.

Changing the Emphasis

Over the past three decades, middle-level educators have come to understand that assessment activities must be realigned. Assessment activities must meet the needs of early adolescents, the middle school organizational pattern of interdisciplinary teaming, and the desired student outcomes of the middle grades curriculum. Rather than assessment driving what we teach and how we teach in the middle-grades, the developmental needs of the student and the scope and sequence of the curriculum must drive assessment. In order to support this claim, educators must strive to develop and implement assessment methods that emphasize learning over memorization and processing skills over test-taking. Authentic assessment allows educators to meet the needs of a curriculum without sacrificing the individual needs of their students. ■

Debunking the Nerd Myth

by Glynis McCray

Glynis McCray *is a research scientist in cell biology at Genentech, Inc., a pioneer biotechnology company. McCray also serves as an advisory board member for the* Math and Science Teacher Education Program (MASTEP). *She has many years of experience as a fund-raiser for science initiative programs and is also a member of the advisory committee for Human Genome Operations, part of the Community College Initiative. She is an active member of ED>NET, a project that facilitates communication between the biotechnology industry and biotechnology programs at community colleges. McCray is a guest lecturer in high school biology for the San Francisco Unified School District and is a motivational speaker for high school and college students. In addition, McCray makes time to serve as a volunteer coordinator for* Access Excellence®, *an on-line professional-development site for biology and life-science teachers. In 1992, she became the co-founder of African Americans in Biotechnology (AAIB), which introduces junior high and high school students to careers in science and biotechnology and provides them with role models.*

The story of my career has many high points. For the past 18 years I have enjoyed working as a scientist in a variety of settings and fields. I have had the opportunity to meet and work with people from around the globe, been given a platform to express my curiosity about how the world works, and have actively contributed to improving the quality of people's lives.

But alas, every story has two sides. The other side of my story begins whenever I meet someone for the first time. The person almost always asks what I do for a living. When I say, "I am a scientist," the response is usually one of anxiety or boredom. A hollow "how interesting" is a common reply. The person then declares an urgent need to make a phone call or makes some other excuse to get away. Needless to say, it is sometimes difficult for me to befriend people who are not scientists or science educators.

Why are people intimidated by what I do? After discussing the phenomena with the few people willing to stick around and listen to me, I found that many people shared some common fears and misconceptions about science and scientists. They thought that science was no fun and too difficult to understand and that scientists were boring, nerdy, or just plain socially inept. After realizing how science and scientists were perceived, I began to see that if my social life was to prosper, I would have to embark on

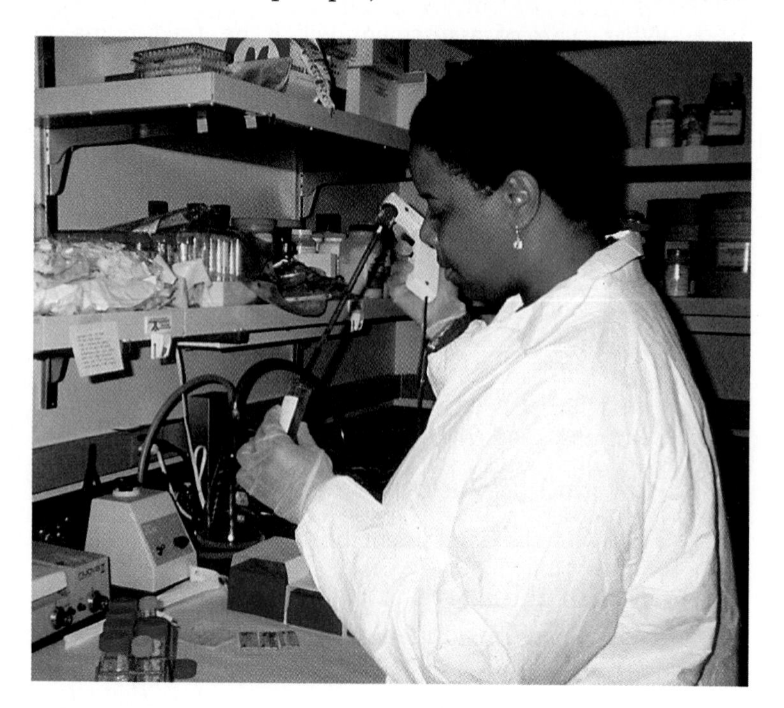

a personal crusade to debunk the myth that scientists are boring, nerdy, and socially inept. Soon, my endeavors to debunk this myth extended beyond the scope of my own social prosperity to a larger desire to make science and the people who do science accessible to a wider range of the population. As a young African-American woman, I no doubt present a non-stereotypical face of a scientist—powerful ammunition for my crusade.

I found that many people shared some common fears and misconceptions about science and scientists.

I decided that my campaign would have three goals. The first would be to give students the opportunity to practice science in a fun and nurturing environment supported by the best teachers and materials I could find. My second goal was to show my students and the community as a whole that a career in science was possible for a diverse group of people. Third, I wanted to help convince educators, parents, and others of the importance of a quality science education for all students and to enlist their help in my crusade.

Debunking in Action

I began at the grass-roots level—at schools in my community. I visited classrooms and talked to students, put on science presentations, and showed students what it means to be a scientist. Also, I began talking with teachers and principals, explaining the goals of my work and enlisting their help. Before long, I expanded my involvement by joining several local and national projects to help develop curricula and activities for the classroom.

To date, my greatest successes, as well as my worst failures, have been with school-age children. At the beginning of my "debunking the myth" campaign, I visited classrooms and science fairs, bringing boundless enthusiasm and science demonstrations that I thought would dazzle and amaze my eager audiences. But in almost every instance, near the halfway point of my demonstration, the students would begin shifting in their seats, yawning, and even falling asleep. Clearly, I was not reaching my audience. And perhaps worse, I was even reinforcing the idea that science and the people who practice it were a bore.

▲ McCray works to debunk the nerd myth among high school students.

Since then, I have discovered that the single most successful approach to getting students interested in the fascinating world of science is to get them involved in the demonstration. By allowing them to take part in the action, I let them discover all on their own how fun and interesting science can be. This is particularly true when the demonstration has real-world connections or some direct relevance to their lives. The benefits to their education and their overall enjoyment are greater still if the experiment can be re-created outside a classroom setting. One of my colleagues still remembers (some 20 years later) a middle school physics experiment about friction. He and his classmates slipped down a playground slide on different surfaces, such as a towel, a rubber mat, and a pair of blue jeans, while another classmate timed the descent. Getting involved in the action like this adds enormous value to the students' experience and can form a lasting impression on them.

Perhaps by doing the experiments themselves, students will begin to comprehend that science is simply a way to understand the world around them and that the people who practice science are folks just like themselves.

One of the most successful hands-on demonstrations I use to get students involved illustrates how the drug Pulmozyme® alleviates a symptom of cystic fibrosis (CF), in which high concentrations of DNA impair airflow in the lungs. After a discussion about CF, students are given two samples of concentrated DNA that simulate the DNA found in the lungs of CF sufferers. A saline solution is added to one of the samples, and a solution containing a DNA-dissolving enzyme is added to the other sample. Drops of each solution are then placed side by side on a microscope slide. The sample with the saline remains viscous, while the sample with the enzyme becomes more fluid and runs down the slide. This activity illustrates how the drug works to loosen DNA and allow more airflow in the lungs of the patient. The hands-on opportunity helps students see how science can lead to developing medicines for people. It also shows students that science and its benefits are understandable and accessible to them.

The single most successful approach to getting students interested in the fascinating world of science is to get them involved in the demonstration.

Science in Our Lives

I believe that the world is currently experiencing one of its most dynamic periods of scientific exploration and discovery. On any given day, newspapers and television feature scientifically significant stories, such as the cloning of animals, breakthroughs in the fight against HIV and cancer, and the discovery of new planets. Current events in science are a powerful tool for positively affecting people's attitudes about science because they show science as it is—dynamic, creative, and inventive. I often use newspaper articles to generate discussions among students and adults on scientific topics as well as to introduce people to interesting areas of science with which they may not be familiar. I also try to speak to people informally about science whenever I perceive an interest. I once spent nearly 2 hours talking to a teenager I met in an in-line skating store about the pros and cons of fat substitutes in potato chips. It was hardly a major victory, but it was another small opportunity to advance the cause.

Expanding the Ranks

Today, the average citizen is asked to understand and make decisions regarding complex political, social, ethical, and economic issues that involve science and technology. An increasingly technology-based workplace requires thinking and reasoning skills that people can learn through science. Clearly, it is more important now than at any other time in our history to have a wide range of people encouraging science education.

The challenge I now face is to find and recruit other scientists, science educators, school administrators, and parents to move the cause forward. To meet this challenge, I communicate with business leaders, community organizers, local school boards, and governments, always with the goal of expanding my circle of fellow debunkers.

I believe that teachers are the most important element in the education of our children. By initiating an outreach program in your community, you can become an advocate for the increased awareness of science education. Outreach programs should include students, fellow teachers, parents, and anyone else interested in the cause. A good start is to create a discussion group to talk about science stories appearing in the daily paper. Through this group, children and adults can share what they know and explore their concerns, each group benefiting from the other's perspective on the issues. Another option for you and your fellow teachers is to create partnerships with local businesses, government offices, and professional organizations that may be willing to provide resources, such as funds, equipment, and volunteers. Without this type of active teacher involvement, there are very few battles we can win. ■

IMPLEMENTING THE NATIONAL SCIENCE STANDARDS

by Juliana Texley, Ph.D.

Juliana Texley *received her doctorate in science education from Wayne State University, in Detroit, Michigan. She has 25 years of experience teaching science at various grade levels, and she was the editor of the National Science Teachers Association's* Science Teacher *for 12 years. She was also the co-editor of NSTA's* Pathways to the Science Standards, *High School Edition, and a contributing editor to the Middle School Edition. Texley is currently Superintendent of Schools for the Anchor Bay School District, in Anchor Bay, Michigan.*

Teachers are used to working with standards. Planning curriculum, defining expectations, and grading students' performance are all part of a a day's work. But in the early 1980s, when a series of more than 300 national studies and reports criticized mathematics and science education in the United States, it was the absence of standards that was most often cited as a major failing.[1] Businesses, governments, and social observers charged that across the nation there was no consensus on what American teachers should teach, how they should teach, or how they should assess students' performance.

Disturbing Headlines During the 1980s, the national education "inferiority complex" made frequent headlines. American students scored at the bottom of international tests. College students in other countries outperformed American

[1] National Committee of Excellence in Education, 1983. (Complete references can be found in *Section III: Continuing the Discussion,* page 144.)

college students in undergraduate course work. Our science, engineering, and technology businesses lacked competent employees. Education experts concluded that before the situation could begin to change, the nation would have to reach a broad agreement on what our school systems should teach students in mathematics and the sciences.

Everyone Has a Say With this challenge, mathematics educators came together relatively quickly to produce a series of new and challenging mileposts by which we could measure a world-class national curriculum.[2] Science educators, on the other hand, took 13 years to come to agreement on a similar set of standards. The path wasn't an easy one. While educators looked for a curriculum model that was broad yet had a depth of content, experts in different fields lobbied hard for their disciplines to be given prominence in the curriculum. Research scientists lobbied for deep content, while educational researchers sought to embed the best inquiry methods into the science standards. Constructivism, the theoretical foundation of much of today's education theory, was being labeled "fuzzy thinking" by many scientists.

Back in the Schools . . . With this flood of input, it was no wonder the first draft of the National Science Education Standards was more than 300 pages long!

But while dozens of national groups argued about the form and substance of the standards, educators began the work.

Even without a national document, science teachers knew the goal—to produce a scientifically literate and economically successful population. Hundreds of science standards projects were generated by school districts, colleges, and other interested groups while the standards committees continued their work.

[2] National Council of Teachers of Mathematics, 1989.

Consensus! Presented by the National Research Council in 1996, the National Science Education Standards codified a consensus on the best in science education. The standards were created by a broad and comprehensive body of contributors and, like the mathematics standards, the science standards were intended to drive and reform both institutions and curricula.

Six Components What set the science standards apart from the mathematics standards was the simultaneous treatment of six distinct components of education. Many earlier standards projects had considered individual or small groups of factors in education. The science standards, on the other hand, acknowledged that in order to create real change, the nation's educators have to consider and address many of the factors influencing education. These factors include *teaching techniques, student assessment, professional development, content, programs,* and *educational systems.*

Practical Value? By addressing our entire educational system, the standards aimed to avoid the lack of coordination that had plagued science curricula in the United States for so long. However, because the standards were written for such a diverse audience and addressed so many different issues, many teachers were left wondering if they would be able to put the standards into practice.

In fact, every teacher has some methods and every classroom has some materials that already reflect the science standards.

The Morning After

The release of the National Science Education Standards made national news. But when the applause died down, it was left to teachers in classrooms to turn the standards into reality. After 13 years of debates and false promises, many science teachers received the final draft of the science standards with mixed emotions—relief that the profession was finally speaking in one voice after so long, and confusion as they wondered, "Now, how do I do it?" Would the hard work and reforms of the past 13 years be discarded? Would teachers be forced to change the way they ran their classrooms? Would teachers be able to respond to the mandate? The task of implementing the science standards, which at first glance seemed simple, suddenly looked very complex.

Transforming the Vision In an effort to respond to teachers' concerns, the NSTA mobilized a large cross-level committee whose purpose was to transform the vision of the standards into practice. The committee's first task was to help science teachers understand what the standards were and what they were not. The committee's purpose was also to link the standards to the hundreds of quality-reform efforts carried out in the years between the publication of *A Nation at Risk: The Imperative for Educational Reform* and the science standards. The project, called *Pathways to the Science Standards,* linked more than 700 existing projects and grant-funded programs with the standards and showed teachers how to move forward from their present practice to the vision of the National Research Council.

A Short Leap In a huge understatement, the authors of the science standards tell teachers that "some outstanding things happen in science classrooms today, even without national standards."[3] In fact, every teacher has some methods and every classroom has some materials that already reflect the science standards. The challenge for teachers is to identify and utilize the practices and materials that are "standards compatible" and to modify current practices to make them even better. The *Pathways* project emphasized that most modern textbook series and most commonly available materials can be used in whole or in part to meet the challenge of the science standards.

---

[3] National Research Council, 1996, page 12.

A Different Angle Science standards challenge teachers to see their own teaching habits and techniques with a new vision. For instance, a favorite hands-on Earth science lab might successfully teach students the names of some rock formations, but it does not encourage students to think about where the formations came from. Consider the differences between two sets of laboratory directions for a familiar experiment in cell biology, below.

LABORATORY DIRECTIONS: BEFORE AND AFTER STANDARDS

Traditional laboratory	Standards-based approach	Links to the standards[4]
We will investigate the efficiency of each cell model to absorb a solution. Iodine will be used as an indicator of starch.	How can we measure how well each cell model absorbs ions? Can you find an indicator that will clearly identify the parts of the starchy potato that are touched?	**Sharing responsibility with the "community of learners"** *Teaching Standard E:* "Enable students to have a significant voice in decisions about the content and context of their work . . ."
Cut a potato into three cubes, 0.5, 1.0, and 1.5 cm on a side.	From a potato, cut three cubes of different sizes, and calculate the surface area and volume of each cube.	**Constructivist discourse and decision making** *Content Standard A:* "As a result of activities in grades 5–8, all students should develop abilities necessary to do scientific inquiry [and] understandings about scientific inquiry." **Integration with mathematics** *Program Standard C:* The K–12 "science program should be coordinated with the mathematics program . . ."
Calculate the percentage of volume absorbed, and place your answer on the chart.	Calculate the percentage of volume absorbed. How accurate is your result? What are the similarities and differences between your model and a real cell?	**Imbedding assessment and self-evaluation** *Assessment Standard C:* "The technical quality of the data . . . is well matched to the decisions and actions taken on the basis of their interpretation." *Teaching Standard B:* "Challenge students to accept and share responsibility for their own learning."
Review your results and prepare for a test.	At home tonight, compare the drying power of an old shirt and a terry cloth towel. How does your work today help you explain your results?	**Unifying Concepts . . . "form and function"** *Content Standard K–12:* "Students should be able to explain function by referring to form and explain form by referring to function."

[4] National Research Council, 1996.

Practical Steps for Moving Toward the Standards

The pathway from traditional hands-on laboratories to more open, standards-based practices is not a long one. Nor is the change particularly difficult, especially if teachers have the confidence to examine their own techniques and the patience to get good results by taking it one step at a time. To move existing teaching practices toward the science standards, consider the following practical steps.

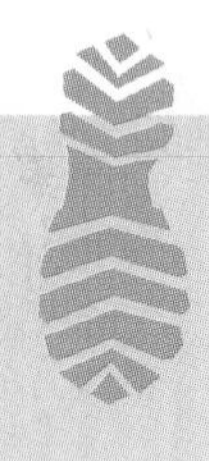

1. Look at your goals for the year.

Narrow the content goals, and add goals in process and skills development. If your district expectations are unrealistic, become an advocate for change.
(*See Teaching Standard A*)

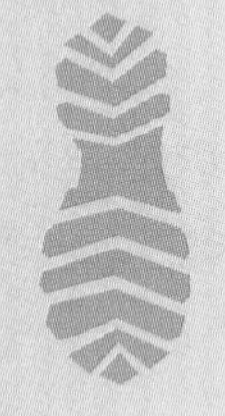

2. Look at last year's lesson plans.

Highlight the exercises or methods that were most successful. They're probably already "up to standard." If you are using a new program this year, find ways to integrate what has worked before into the new materials.
(*See Teaching Standard A*)

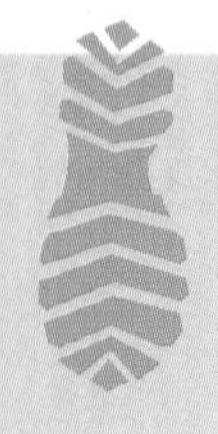

3. Explore your students' preconceptions about new topics before they are introduced.

Use open questioning, journal or brief oral responses, or group dialogues. Open up your discussions to allow students to challenge, question, or simply clarify concepts.
(*See Teaching Standard B*)

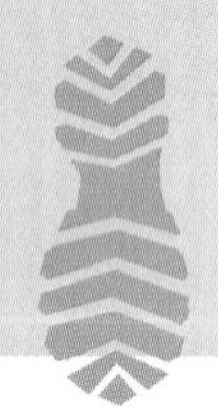

4. Check students' understanding.

Embed assessment into every lesson. Interrupt your normal lesson plan for a written response or group dialogue, or ask students to draw a picture.
(*See Teaching Standard C*)

5. Add a real-world extension to each laboratory or classroom experience.

Add time to your program by developing homework assignments that connect classroom experience to familiar surroundings. Bring parents and the community into the process of extending meaning for students.
(*See Teaching Standard D*)

6. Look at the labs that are most successful for you.

Are there safe ways to let students make their own decisions about parts of the procedure? If so, rewrite those portions. Encourage students to use the methods of scientific research to modify their own work or to develop procedures to answer their own questions. Don't be afraid to explore questions whose answers you do not know. (*See Teaching Standard E*)

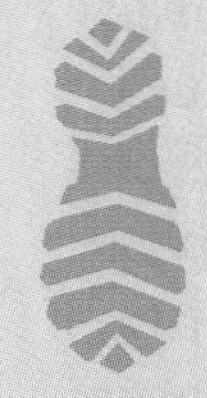

7. Reduce but don't completely eliminate direct, lecture-based instruction.

Include inquiry-based labs in each lesson. When lectures are the most appropriate method of content delivery, find ways to intersperse experiences, quick illustrations, or student-to-student dialogues to involve students in the process and to encourage more-active learning. (*See Teaching Standard E*)

8. Find ways to learn more about your students' differences in background, learning style, or opportunities to learn.

To every successful exercise that you have in your repertoire, add an extension or companion exercise that supports another learning style. For example, add a journal analysis to a good lab or a group discussion to a favorite lecture. (*See Teaching Standard F*)

Beyond the Teaching Methods

The applications of the science standards are not limited to teaching methods. For example, in a school that needs more time for science classes, the standards advocate block scheduling. But when block scheduling isn't possible, the solution may be found in more efficient classroom management or in a teamed, cooperative time-sharing system with a mathematics teacher. If funds for traditional professional development are lacking, the standards support a request for peer coaching or a proposal for action research in classrooms. For the teacher facing an underresourced and overcrowded classroom, the standards address the need for sufficient space and equipment. Although to some the facility guidelines may seem impossible to meet, they can still be useful to teachers in working with administrations to set goals and priorities. For most teachers, the key to reaching the National Science Education Standards will be taking the first few steps and proceeding with confidence.

The Content Standards

As the standards speak to audiences of policy makers and funding agencies, they also give teachers a powerful call to become advocates for their own profession. Probably the most difficult challenge to the teacher-advocate is the call to reduce the amount of content taught in science classes. Despite the standards' foundation in constructivist educational research and the emphasis on inquiry, the standards still emphasize the importance of a strong content base. In fact, one of the most powerful, if controversial, statements made by the authors of the science standards is that students learn science *only* "in the context of content."[5] The content core is narrower and deeper than that of most previous models, but it is still substantive. Because tomorrow's textbooks will probably cover as much content as today's, the standards encourage teachers to choose from the content available to them in their textbooks and to spend relatively more time on the areas they choose. To that extent, the model curriculum reflected in the content standards can be a good support for teachers when they are deciding on content.

The Professional-Development Standards

The professional-development standards demand lifelong support of the nation's educators, placing special emphasis on the content and inquiry methods of research scientists. They also legitimize new and expanded models for professional development, including externships in industry or government agencies and action research in a teacher's own classroom. By making these statements, the standards support new kinds of grant proposals and new uses for professional-development funding, including Eisenhower and Title I funds. The science standards also speak eloquently for equity in education by encouraging every educator to explore, evaluate, and improve students' opportunities to learn within their home and school communities. The standards make a powerful case for compensatory education. By encouraging diverse methods, the standards not only tell teachers why they must enhance equity but also explain how it can be done.

The standards give teachers a powerful call to become advocates for their own profession.

[5] Ibid.

The Program and Systems Standards

The program and systems standards acknowledge that neither teachers nor school systems can implement the science standards alone. For perhaps the first time, there is a high-profile acknowledgment of the responsibility of state and federal educational bureaucracies to become part of the solution rather than part of the problem. For instance, the program standards demand that students have equal access to sufficient resources (Program Standard D) and equitable opportunities to achieve (Program Standard E). The system standards demand that the mandates of systems be supported with resources (System Standard D) and that policies be equitable (System Standard E) and allow for adaptation by individual communities (System Standard A). The system standards also acknowledge that too often in the past what was intended to be a positive change at the state or federal level produced negative effects in classrooms (System Standard F). For example, a new state test intended to increase achievement actually lowered the achievement level of students on objectives that were not tested. The systems standards insist that those who implement policy decisions should involve teachers in the review of those decisions to avoid unforeseen consequences.

The National Science Education Standards aren't radical, but they are unusually realistic.

A New Optimism

By examining the barriers that school systems unintentionally raise to hamper the progress of teachers, the National Science Education Standards have created a new optimism. This is especially true among veteran teachers who have seen the reforms of the 1960s ("The Sputnik Era") and 1980s ("The Back to Basics Movement") come and go. The National Science Education Standards aren't radical, but they are unusually realistic. In time, it will be possible to implement the standards in every classroom and school system. Teachers can move their educational communities forward in small steps by using the standards as mileposts, as support to validate their sound practices, and as ammunition to force systems to recognize the needs teachers have as they progress. There is reason to believe that this reform cycle will produce long-lasting results. ❑

APPLYING FOR EDUCATION GRANTS

by Ernest W. Brewer, Ed.D, and Connie Hollingsworth, Ph.D.

Ernest W. Brewer *and* **Connie Hollingsworth** *have many years' experience writing, implementing, administering, and evaluating grants. Over the past 20 years, they have taught numerous grant writing courses, workshops, and community education classes. Brewer is a professor in and heads the Department of Children and Family Studies at the University of Tennessee, in Knoxville. He has written many articles on grant writing and is best known as the primary author of* Finding Funding: Grant Writing from Start to Finish, Including Project Management and Internet Use. *Hollingsworth is the director of Pre-College Programs at the University of Tennessee.*

Did you hear about the science teacher who was awarded a $20,000 grant to build a rain-forest environment in his classroom? What about the science teacher who received $2,500 to attend a 2-week summer science camp, and even got graduate credit for the experience? Sound exciting? Do you have a good idea for enhancing your teaching or your students' learning but lack the necessary funds for putting the idea into action?

Today, many educators find that in order to provide certain kinds of innovative or hands-on learning experiences for their students, they need to look for funds beyond what the school budget allows. Often, an education grant can be the answer. Writing a grant proposal is not necessarily difficult. But it does require a basic understanding of the process and procedures that are involved. The most important thing to remember is that if you never apply for a grant, you will never get any of the funds available for your program! This article outlines a simple six-step plan that will help you get organized and apply for educational funding.

Six Steps to Funding

1. **Identify your need**
2. **Get pre-approval**
3. **Identify sources of funding**
4. **Contact sources**
5. **Write the proposal**
6. **Submit a letter of intent**

◆ Science summer camp, Austin, TX; funded by NSTA grant and the Weinert Foundation

◆ Girls exploring math and science, Austin, TX; funded by NSTA and Girl Scouts Lone Star Council

1 Identify Your Need

Although you probably have a long wish list of projects and items, it's important to define your needs simply and specifically. Perhaps you need $100 to provide rewards for participants in the school science fair or several thousand dollars to install a computer workstation in your classroom. Or you might be a member of the district's professional-development committee and need money to send teachers to an important conference.

Whatever your needs, limit it to one specific event, item, or activity. For instance, it is better to ask for one set of World Book™ encyclopedias than to ask for 100 new books about science. The first request is simple and specific; the second request is too large for many organizations, and too general. What kinds of books about science? and for whom? Grant donors want to know specifics, and they have a right to know what their money is going to be used for. When formulating an idea for a proposal, try to put yourself in the shoes of the people with the money. What would you want to know about the proposed program before *you* wrote the check? In addition, make sure you can justify your need. You should be able to articulate what you're doing, why you're doing it, and how much money you need. It also means being able to state what the anticipated outcome or benefits will be if your need for funding is met.

2 Get Pre-approval

Before you start working on your grant proposal, be sure to get approval from your immediate supervisor. This may be your principal, superintendent, or other authorized person. First of all, your supervisor may be aware of certain district procedures or requirements that must be met before applying for a grant. Second, administrators in some school districts or schools may have a larger fund-raising strategy of which you are unaware. Organizations and businesses are constantly being asked for contributions and donations. Suppose your assistant superintendent just spent 3 days courting a local soft-drink distributor for a major contribution in excess of $100,000, and a meeting to formalize the donation has been set for 1 week from today. Not knowing this, you and several co-workers start making telephone calls and writing letters to the same company requesting a donation for a smaller project. This unfortunate coincidence would be confusing to the grantor and would reflect badly on you as a potential grantee. An uncoordinated fundraising effort sends a negative message to businesses and organizations and may dissuade them from funding your proposal. Getting approval before beginning to search for funding sources may head off embarrassing or potentially damaging funding conflicts.

Getting approval before beginning to search for funding sources may head off embarrassing or potentially damaging funding conflicts.

3 Identify Sources of Funding

Once you get administrative approval, the next step is to identify sources of funding. There are thousands of grants available to educators. Grants can range from $100 donated by local grocery chains to National Science Foundation awards in excess of $1 million.

Local and Regional Resources

Many community organizations, such as Kiwanis Clubs, Rotary Clubs, and Lions International groups, make small grants (usually under $1,000) available to teachers for innovative education projects. Call the organization, and ask for information about grant applications. A number of retailers, such as Wal-Mart® and J. C. Penney®, allocate funds for community and educational projects as well. These funds are not usually advertised; look for fliers posted near the customer service desk, or call the public relations or customer service department and ask for information.

General Tips for Grant Seekers

- Read professional journals and professional trade magazines. Ads frequently appear in these publications announcing grant competitions.
- Make as many contacts as you can: network, network, network!
- Gather as much background information as you can about the grantor before you begin to write.
- Write to the grantor and request a copy of its most recent annual report and grant guidelines.
- Read and study the information you receive. Target those organizations whose grant outlines most closely match your specific needs.
- Call the organization if you have questions. Ask to speak with the individual who is in charge of grants. Make every effort to get to know this person on a first-name basis.
- Keep a filing system of information about grantors that match your interests and needs. You may not have time to submit a proposal to all of the organizations, but you may find them useful at a later time.
- Set realistic goals for yourself.

State and Federal Resources

Some federally funded agencies, such as the National Science Foundation and the U.S. Department of Education, are major contributors of education grants. Federal grant programs tend to be highly competitive, usually involve money in excess of $100,000, and require extensive paperwork. Awards are made to institutions of higher education (IHEs), state education agencies (SEAs), and local education agencies (LEAs). Federal grants for individual teachers occasionally become available, but these are rare. If you plan to apply as an individual, find grants that are designated for individual teachers and classrooms. However, grant money is still accessible to teachers because the money is often distributed by IHEs, SEAs, and LEAs for individual programs. Your LEA, for example, can submit a grant proposal that designates you as the project director. Applying for large state and federal grants requires a fairly sophisticated grant writing ability. It may be worthwhile to have the proposal written by a grant writing consultant or an individual whose job involves responding to grant notices.

Other Resources

Educators have been highly successful in funding classroom activities and projects by tapping into the resources of private foundations, corporations, and organizations. Good places to begin your search for information about these sources of funding are public libraries, your school district's administrative center (which may have a research or grants office), and local colleges or universities. The Internet is an invaluable place to research sources of funding as well.

Grant notices from private and public organizations often appear in professional journals and trade magazines. Many education honor societies also offer small grants to individuals who are members. Grant information and applications are generally available from the national headquarters of these societies.

4 Contact Sources

Once you have identified and researched four or five organizations with grant programs that seem to match your needs, you are ready to make contact. Before you make your first call, spend some time rehearsing what you plan to say. It might be helpful for you to jot down the important points that you want to include in your introduction. Always have a pen and paper close by so you can write down important names, titles, addresses, telephone numbers, submission dates, and other pertinent information. At the end of the conversation, you should know if this organization's program and your needs are a match. If the organization is not interested in your idea, do not discard your information. Keep it in your files as a reference—you may need to contact this organization in the future.

Tips for Making Contact with Potential Grant Givers

- Be on your best professional behavior by being polite and courteous.
- Avoid using a speaker telephone. Using a speaker telephone is annoying and often appears too impersonal.
- Be brief and to the point. The offices of many charitable institutions are understaffed, and the employees are very busy. They appreciate callers who are clear and concise.
- Find out the name and preferred title of the person you speak with. Keep detailed records of whom you spoke with and when.
- Ask the individual to send you any printed information about his or her organization and its policy for making donations or contributions. Usually this will include some type of brief outline or description that you should follow in requesting funding.
- If you are supposed to call back, be sure to find out the best time to do so. You will save time and avoid appearing pushy or unorganized.
- Thank the individual for allowing you to submit your proposal for consideration, and act quickly.

5 Write the Proposal

So your call has been received positively and the organization invites you to submit a proposal! One of the things that we have noticed in our work with grant writers over the past 20 years is that many individuals get very excited about the prospect of someone giving them grant money, yet very few are willing to follow through and make the effort that it takes to research, plan, organize, and write the proposal. Before you start writing, it is important for you to answer a number of questions.

Pre-Writing Checklist

- ✓ How much time do I have to devote to writing a proposal?
- ✓ Will I be working on the grant alone or with co-workers?
- ✓ Why do I want or need to write a grant proposal?
- ✓ Does my school or school system have guidelines or procedures that I must follow?
- ✓ Am I willing to complete all of the paperwork that may be required by the funding agency if I receive funding?
- ✓ Do I have all the information necessary for writing a grant proposal?

As you write your proposal, keep in mind that everything you write should reflect the "five Cs": Clear, Concise, Cogent, Compelling, and Correct.[1] Keep your purpose clearly in mind as you write. It may be helpful to write your purpose on an index card and refer to the card often as you write. Most agencies will provide you with guidelines about applying for grants. These guidelines may include a simple format to follow for writing your grant proposal.

Overview of a Winning Grant Proposal

I. **Introduction:** Keep this section brief and to the point, usually no more than half a page. Explain your school's location, your educational background and experience, and your current job description.

II. **Description of Your Need:** Be clear and concise. Use facts and figures to back up your claims. Don't paint a tale of woe. Tell the truth, and present information in a fair light.

III. **Goals and Objectives:** The fewer objectives you list, the better. Usually one overarching goal and three to four objectives are sufficient. Your objectives must be measurable! Your benefactors want to know that the donation is sufficient to meet your goal, but is no more than necessary.

IV. **Evaluation Plan:** Assuming you get the grant, an agency will want to know how you will evaluate the success of the project. Be sure to present a carefully thought-out evaluation plan. Let the organization know how and when it can expect to know the outcome of your project and how its donation led to the project's success. The plan doesn't need to contain sophisticated statistical methods, but it does need to reflect sound evaluation principles.

V. **Budget:** A budget should clearly show how the money will be spent. If the agency provides a list of approved expenditures, be sure to include only those items. For instance, it is not uncommon for agencies to specify that no funds be spent on travel expenses. If this is the case, be certain that your budget does not include plane tickets! A common mistake made by novice grant writers is to request too much money or too many donations.

The guidelines also may offer advice on the length of your proposal, mailing instructions, and deadline dates. Most proposals will be from two to five typed pages. All guidelines should be followed carefully.

6 Submit a Letter of Intent

When you submit your proposal, be sure to include a letter of intent. A letter of intent is like a cover letter—it summarizes the reason you are submitting the proposal. Usually, letters of intent are no more than one or two pages and follow a standard format.

A Few Last Words

Don't Forget to Say "Thanks" When you receive a donation or contribution, always thank the donor. Consider inviting a representative to your classroom to see firsthand how the contribution was used or so your students can make personal thank-yous. You could also write an article for your school newspaper or the local newspaper that highlights the donor and the benefits of the donation. Regardless of the follow-up you decide to do, it is always appropriate to send a simple handwritten thank-you note to every individual who helped you obtain the donation.

Dealing with Rejection Getting rejected is never pleasant, and most of us try to avoid it altogether. Unfortunately, in grant writing the only way to avoid rejection is to never apply for a grant in the first place. Assuming that you do apply for a grant, it's important that you learn from rejection. Call or write to the agency and ask how you might improve your proposal for the next competition or review. Some agencies will gladly provide you with this information, greatly improving your chances for success on your next proposal.

A Little Inspiration Billions of dollars are issued each year for grants and contracts. And many of those dollars are granted to individuals who are submitting their first proposals. So put your grant writing fears behind you and start writing!

---

[1] Brewer, et al., 1998. (Complete references can be found in *Section III: Continuing the Discussion,* page 145.)

Sample Letter

January 1, 2001

Sandra Jones
Director of Public Relations
ABC Distributors
111 Money Street
Grantsville, USA 12345-6789

Dear Sandra Jones:

Submitted for your consideration is the proposal that we discussed last week regarding my seventh grade science students at Central Middle School. I am excited about the possibility of having ABC Distributors consider our request to provide $500 toward the purchase of badly needed science equipment for our newly renovated science lab.

Any consideration you can make for this request is greatly appreciated. If you have questions or need further information, please feel free to call me at Central Middle School at 555-6789. I look forward to hearing from you after your board meeting on January 18.

Sincerely,

Andrew Adams

Mr. Andrew Adams
Seventh-Grade Science Teacher

Enclosure

cc: Mrs. Harriet Haley, Principal
Central Middle School

Mr. George Hester, Superintendent
Central City School System

Final Tips from Successful Grant Writers

- Remember that practice makes perfect. Consider each submission part of the learning process, and feel confident in the knowledge that your next grant proposal will be better than your last.
- Talk to experienced grant writers and get their insight into grant writing.
- Take a community education class or college level grant writing class.
- Build a grant library of your own, and read as much as you can about grant writing. There are many excellent resources available in your local library and on the Internet for grant writers.
- Think about your readers as you write, and write to the readers.
- Funding agencies' priorities change, so be sure you check them out carefully before you start writing.
- Always have someone else edit your proposal before you submit it.

Turning an Educator's VISION Into a Classroom REALITY

by Tamra Ivy

Tamra Ivy *teaches middle school science at Albuquerque Academy, in Albuquerque, New Mexico. The project described in this article was developed during the five years she taught eighth-grade Earth science at Ernie Pyle Middle School, in Albuquerque. In addition to classroom teaching, Ivy spent eight years in the Community Education Department of the Albuquerque Public Schools, where she developed and coordinated adult education programs, before- and after-school programs for students, and school-volunteer programs.*

In 1992, dissatisfied with the routine and the results in my eighth-grade Earth science classes, I began to imagine a science classroom where students engaged in their own independent scientific investigations. In this imaginary classroom, students chose their investigation topics, designed and carried out their research plans, and reported their results in a variety of creative and entertaining ways. The students were self-motivated and enthusiastic, responsible and curious, skeptical and open-minded.

> **The students were learning to take notes, but not to think.**

When I shared my vision with colleagues the response was, "Yes, but what about the curriculum you must cover? What if your students are all doing different projects and they all need different lab equipment? How will you monitor such a setup? What about accountability? How can you assess what they are learning? What about tests? What about scope and sequence?" Another point that they emphasized was, "Maybe you could do this with the well-prepared, motivated children who read and write at grade level, but it will never work with the problems we have here."

Still, I knew that my teaching method—a traditional classroom with textbook labs and plenty of note taking—wasn't working. The students were learning to take notes, but not to think. They were learning to follow instructions, but they weren't planning. I needed a structure that would address three major problems faced by my students: low reading ability, high absenteeism, and a limited science background.

The Three Problems at My School

Low reading ability: More than 75 percent of students in our school read at or below the 49th percentile of the Gates-MacGinity reading test. Our school population was approximately 93 percent Hispanic, 6 percent Caucasian, and 1 percent African American and Native American. Because there was no tracking in the science classes, all ranges of language ability were represented in every class. The students' low reading ability prevented us from using our textbooks and other printed teaching materials in the traditional manner.

Absenteeism: Although the reported attendance rate for the school was 95 percent, the actual attendance rate for my eighth graders was 88 percent. Students were absent because they were needed at home; they were participating in a family event; they were ill; or, in some cases, they were skipping school.

Limited science background: The students' background in science was limited mostly by economic factors. Most were from low income families, and often they could not afford books, museum visits, camps, or excursions for the sole purpose of exploring and enjoying science.

Using the Scientific Method

Given these three problems, I knew I would have to individualize the program as much as possible in order to make science relevant for each of my students, but I also knew I needed a strong framework within which to work. I finally decided to focus on the scientific method, specifically as outlined in Gowin's Vee, and have the students apply that method to everything they did.[1] The Vee map is a preprinted format that students use to report their findings. It is a visual organizer for their work (see Figure 1 on page 132 for a sample Vee map[2]). I liked the simple format of the Vee for student lab reports. It was short, focused, and incorporated language development. Students completed the Vee maps in the order that they conducted an investigation.

The chart at the bottom of this page outlines the steps students are to follow when using a Vee map.[3] Originally this was a rather intimidating list with 9 items, so I decided to group the items into a three-step procedure: planning an investigation, doing an investigation, and sharing what you know.

Once I taught the Vee map procedure to the students, I rarely delivered whole-class instruction.

Once I taught the Vee-map procedure to the students, I rarely delivered whole-class instruction. Instead, I taught science skills to individual students in the context of their investigations. I served as an expert in some content areas and referred students to their textbooks for areas outside my expertise. I kept copies of Earth, Life, and Physical science textbooks on hand for students as a reference, and I also referred my students to library materials and other faculty members for additional information.

[1] Gurley-Dilger, L., 1992; W. M. Roth, and M. Bowen, 1993. (Complete references can be found in *Section III: Continuing the Discussion,* page 145.)

[2] The Vee map format pictured in Figure 1, but not its contents, is from: Roth, W. M., and M. Bowen, 1993.

[3] Roth, W. M., and M. Bowen, 1993.

Vee Map Procedure

Planning an investigation:	Doing an investigation:	Sharing what you know:
• Write a focus question. • Read about the topic, and develop a list of vocabulary words. • Make a list of necessary materials. • Describe what the investigation (event) entails.	• Gather data and transform it into an outline, table, graph, or diagram. • Write the knowledge claim that answers the focus question.	• Write a value claim that tells why the new knowledge is important or valuable. • Write a new focus question related to this topic. • Draw a web that explains what you know.

Figure 1: Sample Vee Map
(Items in blue are student generated.)

Name: Josh Riley
Date: September 30
Portfolio #: 4

Focus Question: What is the Bear Canyon habitat like?

Vocabulary:
1. biotic = made of living material
2. abiotic = made of nonliving material

Materials:
1. field notebook
2. pencil

Event:
1. Observe the habitat in Bear Canyon.
2. List what I see, and make sketches.
3. Group observations into categories.

New Focus Question: How does this habitat compare with the other ones I studied?

Data Outline:
I. Animals and Insects
  A. coyotes, bears, squirrels, deer, rabbits, blue jays
  B. caterpillars, crickets, flies, spiders

II. Plants
  A. conifers
    1. ponderosa, Douglas fir, piñon, juniper
  B. deciduous
    1. Gambel oak
  C. shrubs
  D. grasses
  E. flowers

III. Land Forms
  A. mountains, canyons, river ridges

IV. Rocks and Minerals
  A. granite, limestone

V. Climate/Weather
  A. a sample summer day is clear, sunny, breezy, 23°C

Knowledge Claim: Bear Canyon is a mountainous habitat, 1,830–2,745 m in elevation. It supports large animals like bears, deer, and coyotes. The mountain has a granite base and a limestone cap. There is one river in the canyon that provides water. The vegetation in the canyon is varied and plentiful.

Value Claim: This habitat could be in danger. An increasing number of housing developments might threaten the wildlife in Bear Canyon.

Changing the Daily Routine

The greatest help in changing the way I structured my teaching was a book written for language arts teachers, *In the Middle: Writing, Reading and Learning with Adolescents*, by Nancie Atwell.[4] Atwell's highly individualized Writer's Workshop features a dependable routine within which students write and edit their work daily, learning the mechanics of English and increasing their vocabulary and fluency. My goal was to develop a similar format for science in which students learned science skills and content through independent science investigations. I adopted Atwell's routine with almost no variation.

Mini-lessons During mini-lessons, I modeled the methods that I wanted students to develop by conducting my own investigations and sharing the results with them. These mini-lessons became the basis for the only whole-class instruction I delivered, except that the emphasis was on what I learned and how I learned it rather than on presenting predetermined curriculum.

Themes Every few weeks I chose a theme, such as weather, astronomy, or geology. Then during a mini-lesson, the class would brainstorm as many questions as possible about the theme. During that day's status-of-the-class conference, students would give themselves an assignment and then proceed through the set of steps described earlier.

Misconceptions If a student wanted to do an investigation that compared the characteristics of the seven planets in the solar system, I learned to bite my tongue and let the student proceed. Misconceptions served as the best starting point for investigations. Often, students' questions came from their own experience and reflected things that they and their families were concerned about.

Ownership Students' ownership of their topics and their work made classroom management easy because the students were self-motivated. However, being eighth-graders, their biggest problem was wasting time or working too slowly. The average investigation took 3–5 days to complete: a day for planning, 1–3 days for gathering data, and a day for preparing the report. The status-of-the-class conference worked like a roll call; students told me which steps they were working on when I called their names. If a student spent too much time on one step, we conferred to find out why he or she was not making progress. Those students who finished before the others selected a new topic within the theme.

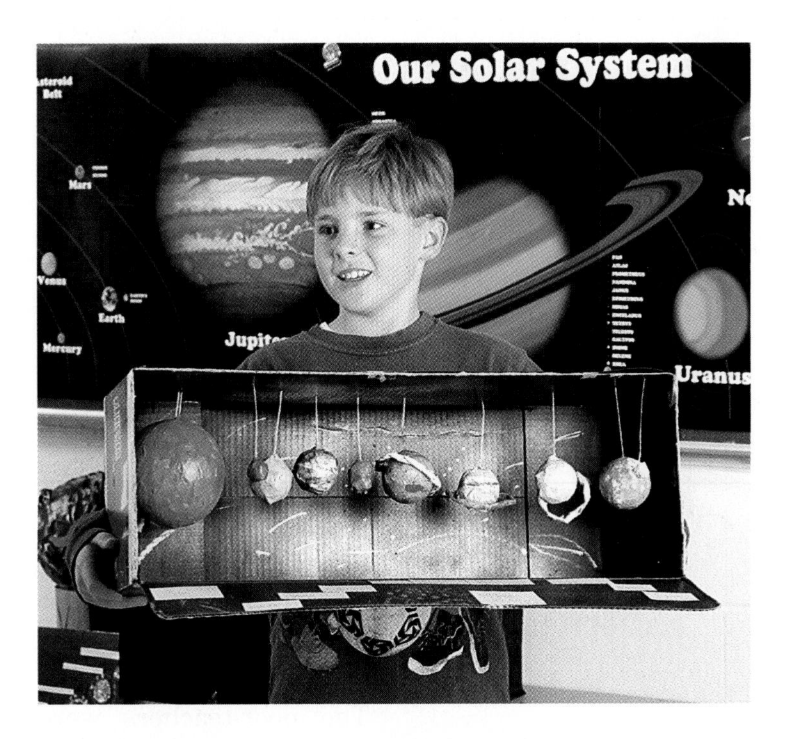

Daily Science Class Routine

Review *(5 minutes)*—Review of work from the previous day; drill on the pooled vocabulary from all current investigations; and review of the previous day's mini-lesson.

Mini-lesson *(10 minutes)*—New information presented about a current theme or specific skills. Topics drawn from a variety of sources, such as the textbook, news or magazine articles, or reference materials. Students usually take brief notes.

Status-of-the-class conference *(5 minutes)*—Roll call of student progress.

Science workshop *(20 minutes)*—Students conduct their investigations independently or in groups. They plan new investigations, carry out current investigations, or prepare their Vee maps.

Wrap-up *(5 minutes)*—Students write in their notebooks, reflect on what they accomplished, and make reminders for the following day's work.

[4] Atwell, N., 1987. (The new title is *In the Middle: New Understandings About Writing, Reading, and Learning*, 1998.)

Teaching Strategies

One of the first things I learned was that I could not stay in front of the class. I wandered through the lab, ready to answer technical questions, prod students along, pose questions, settle arguments, or listen to eighth-grade tales of woe.

> Gradually, I became a tutor, a coach, a cheerleader, and a facilitator.

Peer Teachers Because I could not be everywhere in the class at once, I trained and relied on peer-teachers to assist students who were having difficulty with the procedures, the content they were studying, or the English language. The peer-teachers were other students who were proficient in the steps of scientific investigation. They were usually, but not always, students who finished work well ahead of time and who liked working with other students. They documented the goals of their students and recorded their progress. Most of the time, the peer-teachers had extremely high goals for their students and coached them into A-quality work. Eventually the peer-teachers could manage the classroom, quiz other students, and give whole-class demonstrations.

My Role Gradually, I became a tutor, a coach, a cheerleader, and a facilitator. Some days I was merely a "gofer," collecting and distributing lab materials. Other days, the students didn't seem to need me there at all. Throughout all the changes, however, it remained my job to assess student work; this gave me the most anxiety. Because my approach was so experimental, it didn't seem right to administer standardized tests. Nevertheless, I needed some form of evaluation and feedback for the students and for myself.

Comprehensive Assessment

How do you assess student work when all of your students are working on different projects? How do you know how well they are doing relative to building, district, state, or national norms? How do you know if you are covering the curriculum? Again, part of the answer came from Nancie Atwell. I adapted the status-of-the-class conference, quarterly grade conferences, and pre- and post-surveys to a science lab format. As I described earlier, the status-of-the-class conference provided a way to monitor student work on a daily basis. At the end of an investigation, student work was placed in a portfolio, which also became part of the evaluation process.

Quarterly Conferences During the quarterly grade conferences, students evaluated their own work and set new goals for themselves. I based their grades on how well they completed their self-assigned work at the predetermined level of 80 percent mastery: the Vee maps had to be complete, and they had to make sense. I required that the parts be interrelated and that the knowledge claim had to answer the focus question. I also used these quarterly conferences

to encourage the students to attempt more sophisticated and in-depth investigations.

Surveys Pre- and post-surveys served as final exams. Students compared their end-of-the-year answers with their beginning-of-the-year answers and analyzed why their answers had changed. These instruments revealed the students' gains in maturity, self-motivation, and growth in their understanding of the scientific process. Surveys do not, of course, test scientific knowledge. For that, I had to develop other assessment methods.

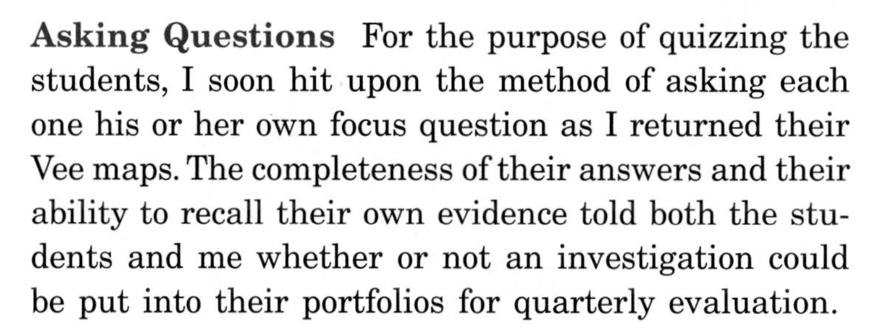

Asking Questions For the purpose of quizzing the students, I soon hit upon the method of asking each one his or her own focus question as I returned their Vee maps. The completeness of their answers and their ability to recall their own evidence told both the students and me whether or not an investigation could be put into their portfolios for quarterly evaluation.

Performance-Based Assessment By mid-year, I began experimenting with performance-based assessment methods. My goal was to assess students' application and analysis skills. I allowed students to work in groups to solve a problem that I felt encompassed most of what they had worked on individually. The result was that they pooled their individual knowledge, learned from each other, and worked together to solve problems.

The Iowa Test of Basic Skills (ITBS) served as the only quantitative measurement device that I used. It was a challenge to let go of all standardized tests except the ITBS, but I began to develop a richer and more comprehensive approach to evaluation than standardized tests.

Documenting Progress

More Investigations With help from my school district's Title II office, I was able to track and document student achievement and results throughout the year. First, and most surprising to me, was the number of investigations students completed. The approximately 140 students in six classes finished 2,823 independent scientific investigations during the 1992–1993 school year, an average of over 20 per student. Because of the individualized nature of the program, students did not lose instruction when they were absent. The Vee maps were portable, and students carried out investigations while they were hunting and fishing, visiting in Mexico, recovering from the chicken pox, and waiting out suspensions.

Students pooled their knowledge, learned from each other, and worked together to solve problems.

Better Attitudes Second, and less obvious, was the students' change in attitude toward science and their ability to do science. At the beginning of the year, students believed they needed special tools and advanced degrees to do scientific investigations; by the end of the year, they reported that they needed good topics, curiosity, good data, and the ability to do work. By the end of the year, most students expressed great confidence in their ability to carry out scientific investigations. (Please see Figure 2 on the next page.)

Figure 2. Student Answers from Pre- and Post-Class Surveys			
		Pre-	**Post-**
1. Are you a scientific investigator?	Yes No Other	9% 90 1	70% 30 0
2a. How did you learn to be a scientific investigator?	Schooling Doing Don't know	66% 17 17	53% 46 1
2b. How do other people learn to be scientific investigators?	Schooling Doing Don't know	53% 28 19	47% 47 6
3. Why do people do scientific investigations?	To find things out Don't know	88% 12	100% 0
4. What do people need in order to do scientific investigations?	External things Information Internal things Don't know	36% 23 10 31	16% 41 43 0
5. How does your teacher decide which investigations are good ones?	Quality of work Learning Other Don't know	11% 4 30 55	5% 23 14 7
6. In general, how do you feel about what you investigate?	Good Ambivalent Don't know Not so good	23% 11 58 2	75% 14 6 5

Better Scores Third, and to my great relief, their science scores on the Iowa Test of Basic Skills rose from an average of 30 in 1992 to 34 in 1993, and science was the highest score for the entire battery. In fact, for many students science was now their best subject. Although the rise cannot be attributed solely to the change in teaching, I was now assured that this method did not harm them and that they had learned more than the previous year's students.

> **Their ability to communicate improves because they, like us, want to talk about the work they are doing . . .**

What I Learned

Although I didn't quite attain my classroom vision, I came close. I learned that, yes, eighth-graders can conduct independent investigations—more than you will want to grade—and that they can become responsible for their own work. When students do assume responsibility, their self-esteem rises because they feel that they are competent, and indeed they are. Their ability to communicate improves because they, like us, want to talk about the work they are doing, and they move from being externally controlled and motivated to being internally controlled and motivated. Moreover, the process of investigating discrepancies helped students develop language and higher-order thinking skills. As the year progressed, their Vee maps became fuller and more complete because phrases turned into sentences and sentences became paragraphs.

I also learned that my students could make good use of reading materials for their projects even if they were not able to tackle long reading or writing exercises. My students used their textbooks as an important reference tool and support for their investigations, but they were not inclined to organize their own thoughts and the sequence of their lessons around the book's organization. In addition, I found that my students were more inclined to challenge themselves to work through a piece of writing that was at or above their reading level when they were investigating a topic of their choosing.

And finally, I learned that I cannot return to a teacher-centered classroom, in which I control the content, orchestrate the labs, and administer the tests. It seems more important to begin with each student, wherever he or she is, and go forward from there.

In fact, for many students science was now their best subject.

Making Science Accessible

This method requires great faith in your students' abilities to learn. They will rise or fall with your expectations. As we are reminded in *Science for All Americans:*

> *Students respond to their own expectations of what they can and cannot learn. If they believe they are able to learn something, whether solving equations or riding a bicycle, they usually make headway.*[5]

Moreover, students are quick to pick up the expectations of success or failure that others have for them. The positive and negative expectations shown by parents, counselors, principals, peers, and—more generally—by the news media affect students' expectations and hence their learning behavior.[6] ■

---

[5] Rutherford, F. J., and A. Ahlgren, 1990, page 187.

[6] Ibid.

Continuing the Discussion

Contents

Footnotes and Additional Resources for Professional Articles

More Resources!

Footnotes and Additional Resources for Professional Articles

The Well-Managed Classroom

Harry K. Wong
pages 34–39

➲ Footnoted References

Sanford, Julie, et al. "Improving Classroom Management." *Educational Leadership,* vol. 40, no. 7, (April 1983), 56–60.

Wong, Harry K., and Rosemary T. Wong. *The First Days of School: How to Be an Effective Teacher* (rev. ed.). Mountain View, CA: Harry K. Wong Publications, 1998.

➲ Additional Resources

Burnard, Sonia. *Developing Children's Behaviour in the Classroom: A Practical Guide for Teachers and Students.* Washington, DC: Falmer Press, 1998.

Williams, Patricia. *Manage Secondary Classrooms: Principles and Strategies for Effective Management and Instruction.* Boston: Allyn and Bacon, 1999.

Wilson, Liz, ed. *Creating an Orderly Learning Environment.* ERS Information Folio Series. Arlington, VA: Educational Research Service, 1998.

The Top 10 Things New Teachers Should Know

Catherine Wilcoxson, Ph.D.
pages 40–43

➲ Additional Resources

Brock, Barbara L., and Marilyn L. Grady. *From First-Year to First-Rate: Principals Guiding Beginning Teachers.* Thousand Oaks, CA: Corwin Press, 1997.

Enz, Billie. *Teacher's Toolbox: A Primer for New Professionals.* Dubuque, IA: Kendall/Hunt Publishing Co., 1997.

Hopkins, David, et al. *Creating the Conditions for Classroom Improvement: A Handbook of Staff Development Activities.* Bristol, PA: Taylor and Francis, 1997.

Kottler, Ellen. *Secrets for Secondary School Teachers: How to Succeed in Your First Year.* Thousand Oaks, CA: Corwin Press, 1998.

Kronowitz, Ellen L. *Your First Year of Teaching and Beyond* (3d ed.). New York: Longman Publishing Group, 1999.

MacDonald, Robert E. *A Handbook of Basic Skills and Strategies for Beginner Teachers: Facing the Challenge of Teaching in Today's Schools.* New York: Addison Wesley Longman, 1999.

Murray, Barbara A., and Kenneth T. Murray. *Pitfalls and Potholes: A Checklist for Avoiding Common Mistakes of Beginning Teachers.* West Haven, CT: National Education Association Professional Library, 1997.

Phelps, Patricia H. *Beginning to Teach: Strategies for a Successful First Day.* Beginning to Teach Booklet Series. West Lafayette, IN: Kappa Delta Pi, 1998.

Portner, Hal. *Mentoring New Teachers.* Thousand Oaks, CA: Corwin Press, 1998.

Ramage, Katherine, and Mark St. John. *The Influence of Professional Development on Classroom Practice: The California Reading and Literature Project.* Studies of the California Subject Matter Projects, Report 12. Inverness, CA: Inverness Research Associates, 1997.

Thurston, Cheryl M., ed. *Survival Tips for New Teachers: From Teachers Who Have Been There (and Lived to Tell About It).* Fort Collins, CO: Cottonwood Press, 1997.

Yes, Teaching Students to Argue Is a Good Idea . . . No, I'm Not Crazy!

Valerie Goff Whitecap
pages 44–49

➲ Footnoted References

Rancer, Andrew S., Valerie G. Whitecap, et al. "Testing the Efficacy of a Communication Program to Increase Argumentativeness and Argumentative Behavior in Adolescents." *Communication Education,* vol. 46 (1997), 273–286.

➲ Additional Resources

Infante, D. A. *Arguing Constructively.* Prospect Heights, IL: Waveland Press, 1988.

Lewis, Barbara A. *The Kid's Guide to Social Action: How to Solve the Social Problems You Choose—and Turn Creative Thinking Into Positive Action.* Minneapolis: Free Spirit Publishing, 1998.

Lewis, Barbara A. *What Do You Stand For? A Kid's Guide to Building Character.* Minneapolis: Free Spirit Publishing, 1997.

Understanding Aggressive Communication

Andrew S. Rancer, Ph.D.
pages 50–53

➲ Footnoted References

Infante, D. A. *Arguing Constructively.* Prospect Heights, IL: Waveland Press, 1988.

Infante, D. A. "Teaching Students to Understand and Control Verbal Aggression." *Communication Education,* vol. 44 (1995), 51–63.

Infante, D. A., and A. S. Rancer. "A Conceptualization and Measure of Argumentativeness." *Journal of Personality Assessment,* vol. 46 (1982), 72–80.

Rancer, Andrew S., et al. "Testing the Efficacy of a Communication Program to Increase Argumentativeness and Argumentative Behavior in Adolescents." *Communication Education,* vol. 46 (1997), 273–286.

Infante, D. A., A. S. Rancer, and D. F. Womack. *Building Communication Theory* (3rd ed.). Prospect Heights, IL: Waveland Press, 1996.

➲ Additional Resources

Infante, D. A., and C. J. Wigley. "Verbal Aggressiveness: An Interpersonal Model and Measure." *Communication Monographs,* vol. 53 (1986), 61–69.

Strategies for Improving Student Behavior

M. Lee Manning, Ph.D.
pages 54–59

➲ Footnoted References

Brophy, J. E. "Classroom Organization and Management." *The Elementary School Journal,* vol. 83, no. 4 (1983), 265–285.

Brophy, J. E., and T. L. Good. "Teacher Behavior and Student Achievement." In *Handbook of Research on Teaching* (3d ed.), edited by M. C. Wittock, 328–375. New York: Macmillan, 1986.

Doyle, W. "Classroom Organization and Management." In *Handbook of Research on Teaching* (3d ed.), edited by M. C. Wittock, 392–431. New York: Macmillan, 1986.

Good, T. L., and J. E. Brophy. *Looking in Classrooms* (6th ed.). New York: HarperCollins, 1994.

Porter, A. C., and J. E. Brophy. "Synthesis of Research on Good Teaching: Insights from the Work of the Institute of Research on Teaching." *Educational Leadership,* vol. 45, no. 8 (1988), 74–85.

Walberg, H. J. "Synthesis of Research on Time and Learning." *Educational Leadership,* vol. 45, no. 6 (1988), 76–85.

➲ Additional Resources

Burnard, Sonia. *Developing Children's Behaviour in the Classroom: A Practical Guide for Teachers and Students.* Washington, DC: Falmer Press, 1998.

Gordon, Gerard. *Managing Challenging Children.* Greenwood, WA: Prim-Ed, 1996.

Lavelle, Lynn. *Practical Charts for Managing Behavior.* Austin, TX: Pro-Ed, 1998.

Rangasamy, Ramasamy. "Reduction of a Classroom Behavior Problem Through a Multiple-Element Intervention." *Education,* vol. 115, no. 1 (1994).

Watson, George. *Classroom Discipline Problem Solver: Ready-to-Use Techniques and Materials for Managing all Kinds of Behavior Problems.* Des Moines: Center for Applied Research in Education, 1998.

West, Richard P., et al. "The Musical Clocklight: Encouraging Positive Classroom Behavior." *Teaching Exceptional Children,* vol. 27, no. 2 (1995).

Wolfgang, Charles H., and Judith L. Irvin. *Strategies for Teaching Self-Discipline in the Middle Grades.* Old Tappan, NJ: Allyn and Bacon, 1999.

Motivate the Unmotivated with Scientific Discrepant Events

Emmett L. Wright, Ph.D.
pages 60–65

➲ Additional Resources

Callison, Priscilla, and Emmett L. Wright. "Huffin' and Puffin' Your Way Through Science." *Science Activities,* vol. 34, no. 1 (1997), 29–32.

Govindarajan, Girish, and Emmett L. Wright. "How Stimulating Ideas Can Generate An Attitude of Inquiry." *Journal of Science Education,* vol. 1, no. 8 (1996), 71–77.

Govindarajan, Girish, and Emmett L. Wright. "Using Minds-On Scientific Discrepant Events to Motivate Disinterested Science Students Worldwide." *Science Education International,* vol. 5, no. 2 (1994), pages 17–21.

Haven, Kendall. *Marvels of Science: 50 Fascinating 5-Minute Reads.* Englewood, CO: Libraries Unlimited, 1994.

The Wild Goose. *Science Without Answers.* Salt Lake City, UT: Wild Goose Company, 1997.

Wlodkowski, Raymond J., and Judith H. Jaynes. *Eager to Learn: Helping Children Become Motivated and Love Learning.* San Francisco: Jossey-Bass, 1991.

Wright, Emmett L., and Girish Govindarajan. "Stirring the Biology Teaching Pot with Discrepant Events." *The American Biology Teacher,* vol. 54, no. 4 (1992), 205–210.

Wright, Emmett L., and Girish Govindarajan. *Teaching with Scientific Conceptual Discrepancies: Unlocking the Mind to Problem Solving.* Manhattan, KS: College of Education (Kansas State University), 1994.

Ensuring Girls' Success in Science

Mary Sandy
pages 66–71

Footnoted References

Burger, Carol, and Mary Sandy. *A Guide to Gender Fair Education in Science and Mathematics.* Charleston, WV: Eisenhower Regional Consortium for Mathematics and Science Education at the Appalachia Educational Laboratory, 1998.

Sadker, Myra, and David Sadker. *Failing at Fairness: How America's Schools Cheat Girls.* New York: Charles Scribner's Sons, 1994.

Skolnick, Joan, et al. *How to Encourage Girls in Math and Science: Strategies for Parents and Educators.* Palo Alto, CA: Dale Seymour Publications, 1982.

The Wellesley College Center for Research on Women. *The AAUW Report: How Schools Shortchange Girls,* by Susan Bailey, et al. Washington, DC: AAUW Educational Foundation, 1992.

Additional Resources

Campbell, Patricia B., and Jennifer N. Storo. *Teacher Strategies That Work for Girls and Boys.* Washington, DC: U.S. Dept. of Education, Office of Educational Research and Improvement, Educational Resources Information Center, 1996.

Hede, Karyn. "Single-Sex Classrooms: Girls, Math and Science." *Scientific American Presents,* vol. 9, no. 2, 1998.

National Science Foundation. *Woman and Minorities in Science and Engineering.* Arlington, VA: National Science Foundation, 1994.

Rhoton, Jack, and Patricia Bowers, eds. *Issues in Science Education.* Arlington, VA: National Science Teachers Association, 1996.

Samuels, Linda S. *Girls Can Succeed in Science! Antidotes for Science Phobia in Boys and Girls.* Thousand Oaks, CA: Corwin Press, 1999.

Teaching Science to Students with Limited English Proficiency

Berty Segal Cook
pages 72–77

Additional Resources

Ballenger, Cynthia. "Social Identities, Moral Narratives, Scientific Argumentation: Science Talk in a Bilingual Classroom." *Language and Education,* vol. 11, no. 1 (1997).

Jasko, Susan A. "Education and Assessment: How Do We Measure a Game?" *Education,* vol. 118, no. 1 (1997).

Markham, Paul, et al. "Identification of Stressors and Coping Strategies of ESL/Bilingual, Special Education, and Regular Education Teachers." *The Modern Language Journal,* vol. 80, no. 2 (1996).

Segal-Cook, Bertha E. *Practical Guide for the Bilingual Classroom* (Spanish rev. ed.). Brea, CA: Berty Segal, Inc., 1994.

Segal-Cook, Bertha E. *Teaching English Through Action* (9th ed.). Brea, CA: Berty Segal, Inc., 1999.

Sheffield, Caryl J. "Instructional Technology for Teachers: Preparation for Classroom Diversity." *Educational Technology,* vol. 37, no. 2 (1997).

Verma, Mahendra K., et al., eds. *Working with Bilingual Children: Good Practice in the Primary Classroom.* Bilingual Education and Bilingualism Series, vol. 6, 1995.

Meeting the Needs of the Academically Gifted

Sally M. Lafferty, Ed.D.
pages 78–83

Footnoted References

Clickenbeard, P. R. "Unfair Expectations: A Pilot Study of Middle School Students' Comparisons of Gifted and Regular Students." *Journal for the Education of the Gifted,* vol. 15, no. 1 (1991), 56–63.

George, P. "Talent Development and Grouping in the Middle Grades: Challenging the Brightest Without Sacrificing the Rest." *Middle School Journal,* vol. 15, no. 4 (1995), 12–17.

Kaplan, Sandra. Keynote Address, Utah Association for Gifted Students, January 1998. Salt Lake City, UT.

Kulic, J. A., and L. C. Kulic. "Effects of Ability Grouping on Student Achievement." *Equity and Excellence,* vol. 23, nos. 1 and 2 (1987), 22–30.

National Middle School Association. *This We Believe: Developmentally Responsive Middle Level Schools.* Columbus, OH: National Middle School Association, 1995.

Tomlinson, Carol Ann. "Deciding to Differentiate Instruction in Middle School: One School's Journey." *Gifted Child Quarterly,* vol. 39, no. 2 (1995), 77–87.

Tomlinson, Carol Ann. "Gifted Learners: The Boomerang Kids of the Middle School?" *Roeper Review,* vol. 16, no. 3 (1994), 177–182.

➲ Additional Resources

Rimm, Sylvia B. *Why Bright Kids Get Poor Grades: and What You Can Do About It* (2d ed.). New York: Crown Books for Young Readers, 1996.

Rogers, Karen B. *The Relationship of Grouping Practices to the Education of the Gifted and Talented Learner: Research-Based Decision Making Series.* Storrs, CT: National Research Center on the Gifted and Talented, 1991.

Schmitz, Connie C., and Judy Galbraith. *Managing the Social and Emotional Needs of the Gifted: A Teacher's Survival Guide.* Minneapolis: Free Spirit Publishing, 1985.

Tomlinson, Carol Ann. *Differentiating Instruction for Mixed-Ability Classrooms.* Alexandria, VA: Association for Supervision and Curriculum Development, 1996.

Winebrenner, Susan. *Cluster Grouping of Gifted Students: How to Provide Full-time Services on a Part-time Budget.* Washington, DC: U.S. Dept. of Education, Office of Educational Research and Improvement, Educational Resources Information Center, 1993.

Winebrenner, Susan. *Teaching Gifted Kids in the Regular Classroom: Strategies and Techniques Every Teacher Can Use to Meet the Academic Needs of the Gifted and Talented.* Minneapolis: Free Spirit Publishing, 1992.

Winebrenner, Susan. *Teaching Kids with Learning Difficulties in the Regular Classroom: Strategies and Techniques Every Teacher Can Use to Challenge and Motivate Struggling Students.* Minneapolis: Free Spirit Publishing, 1996.

Making Hands-on Doable

John G. Upham
pages 84–87

➲ Additional Resources

Brown, Janet Harley. *Assessing Hands-on Science: A Teacher's Guide to Performance Assessment.* Thousand Oaks, CA: Corwin Press, 1996.

Hauser, Jill Frankel. *Super Science Conconctions: 50 Mysterious Mixtures for Fabulous Fun.* Charlotte, VT: Williamson Publishing, 1996.

Science Education Suppliers Guide. Annual publication. Washington, DC: National Science Teachers Association, 1999.

Shevick, Ed. *Science Action Labs.* Woodland Hills, CA: Ed Shevick, 1994–1997.

Tolman, Marvin N. *Hands-on Life Science Activities for Grades K–8.* West Nyack, NY: Parker Publishing Company, 1996.

Tolman, Marvin N. *Hands-on Physical Science Activities for Grades K–8.* West Nyack, NY: Parker Publishing Company, 1995.

The Internet: Realizing the Potential

David Warlick
pages 88–93

➲ Footnoted References

Cohen, Moshe, and Margaret Riel. "The Effect of Distant Audiences on Students' Writing." *American Educational Research Journal,* vol. 26, no. 2 (1989), 143–159.

➲ Additional Resources

Association for Supervision and Curriculum Development. *Learning with Technology: 1998 ASCD Yearbook.* Alexandria, VA: Association for Supervision and Curriculum Development, 1998.

Bennett, M., and K. Walsh. "Desperately Seeking Diversity: Going Online to Achieve a Racially Balanced Classroom." *Computers and Composition,* vol. 14, no. 2 (1997).

Bozzone, Meg A. "Schools That Work: Technology for Kids' Desktops—How One School Brought Its Computers out of the Lab and into Classrooms." *Electronic Learning,* vol. 16, no. 5 (1997).

Britton, Zachary. *Safety Net: Guiding and Guarding Your Children on the Internet.* Eugene, OR: Harvest House Publishers, 1998.

Corbett, Angela. "Unleashing the Power of the Internet as a Classroom Learning Tool." *Computer Education.* vol. 40, no. 85 (1997).

Garrett, Alan W. "Computers, Curriculum, and Classrooms: Panacea or Patent Medicine?" *Journal of Curriculum and Supervision,* vol. 13, no. 1 (1997).

Koch, Melissa, and Ferdi Serim. *NetLearning: Why Teachers Use the Internet,* Songlines Guides Series. Sebastopol, CA: O'Reilly and Associates, 1996.

McGreal, Roy. "The Internet: A Learning Environment." *New Directions for Teaching and Learning,* no. 71 (Fall 1997).

McLain, Timothy, et al. *Educator's Internet Companion: Classroom Connect's Complete Guide to Resources on the Internet.* Lancaster, PA: Wentworth Worldwide Media, 1995.

Riley, Richard W. "Connecting Classrooms, Computers, and Communities." *Issues in Science and Technology,* vol. 12, no. 2 (1996).

Roerden, Laura Parker. *Net Lessons: Web-Based Projects for Your Classroom.* Sebastopol, CA: Songline Studios, and O'Reilly and Associates, 1997.

Royer, Regina. "Teaching on the Internet—Creating a Collaborative Project." *Learning and Leading with Technology,* vol. 25, no. 3 (1997).

Organizing a Family Science Night at Your School

Lloyd H. Barrow, Ph.D.
pages 94–99

➲ Additional Resources

Gardner, Robert. *Science Projects About Kitchen Chemistry.* Springfield, NJ: Enslow Publishers, 1999.

Haslam, Andrew. *World Book of Science Fairs: Ideas and Activities.* Chicago: World Book, 1998.

Herbert, Don. *Mr. Wizard's 400 Experiments in Science.* North Bergen, NJ: Book-Lab, 1983.

Herbert, Don. *Mr. Wizard's Experiments for Young Scientists.* New York: Doubleday, 1990.

Herbert, Don. *Mr. Wizard's Supermarket Science.* New York: Random House, 1980.

Murphy, Pat, et al. *The Science Explorer: Out and About.* Exploratorium Science-at-Home Series. New York: Henry Holt, 1995.

Teaching and Learning Science Through Writing

Carol M. Santa, Ph.D., and Lynn T. Havens
pages 100–105

➲ Footnoted References

Calkins, L. *The Art of Teaching Writing.* Portsmouth, NH: Heinemann, 1987.

Danscreau, D. F., "Learning Strategy Research." In *Thinking and Learning Skills,* edited by J. W. Segal, S. F. Chipman, and R. Glaser, 209–240. Hillsdale, NJ: Erlbaum, 1985.

Palinesar, A. S., and A. L. Brown. "Interactive Teaching to Promote Independent Learning from Text." *The Reading Teacher,* vol. 39 (1986), 771–777.

Santa, C. M., and L. T. Havens. "Learning Through Writing." *Science Learning: Processes and Applications,* edited by C. M. Santa and D. E. Alvermann. Newark, DE.: International Reading Association, 1991.

➲ Additional Resources

Atwell, Nancie. *In the Middle: Writing, Reading, and Learning with Adolescents.* Portsmouth, NH: Boynton/Cook Publishers, 1987.

Goldberg, Howard, and Philip Wagreich. "Focus on Integrating Science and Math." *Science and Children*, vol. 26, no. 5 (1989).

Graves, D. *Writing: Teachers and Children at Work.* Portsmouth, NH: Heinemann, 1983.

Santa, Carol M., and Donna E. Alverman, eds. *Science Learning: Processes and Applications.* Ann Arbor, MI: Books on Demand, 1991.

Tierney, Bob. *How to Write to Learn Science.* Arlington, VA: National Science Teachers Association, 1996.

Assessment That Emphasizes Learning

Sandra L. Schurr, Ph.D.
pages 106–111

➲ Footnoted References

Bloom, B. *Taxonomy of Educational Objectives.* New York, NY: David McKay Co., Inc., 1956.

➲ Additional Resources

Barber, Jacqueline, et al. *Insights and Outcomes: Assessments for Great Explorations in Math and Science.* Great Explorations in Math and Science (GEMS) Series. Berkeley, CA: University of California, Berkeley, Lawrence Hall of Science, 1995.

Combs, Dorie. "Using Alternative Assessment to Provide Options for Student Success." *Middle School Journal,* vol. 29, no. 1 (September 1997).

Doran, Rodney, et al. *Science Educator's Guide to Assessment.* Arlington, VA: National Science Teachers Association, 1998.

Schurr, S., and J. Thomason. *Teaching at the Middle Level: A Professional's Handbook,* edited by M. Thompson and J. H. Lounsbury. Lexington, MA: D. C. Heath and Co., 1996.

Forte, I., and S. Schurr. *Authentic Assessment. The A–Z Active Learning Series.* Nashville, TN: Incentive Publications, 1997.

Forte, I., and S. Schurr. *The Definitive Middle School Guide.* Nashville, TN: Incentive Publications, 1993.

Forte, I., and S. Schurr. *Making Portfolios, Products, and Performances Meaningful and Manageable for Students and Teachers: Instructional Strategies and Thematic Activities.* Nashville, TN: Incentive Publications, 1995.

National Education Goals Panel. *The National Education Goals Report: Executive Summary: Commonly Asked Questions About Standards and Assessment.* Upland, PA: Diane Publishing Company, 1997.

Pallrand, George J. "The Relationship of Assessment to Knowledge Development in Science Education." *Phi Delta Kappan,* vol. 78, no. 4 (1996).

Rezba, Richard J., Constance Sprague, Ronald L. Fiel, and James H. Funk. *Learning and Assessing Science Process Skills* (3d ed.). Dubuque, IA: Kendall/Hunt Publishing Co., 1995.

Tierney, Robert J., Mark Carter, and Laurie Desai. *Portfolio Assessment in the Reading-Writing Classroom.* Norwood, MA: Christopher-Gordon Publishers, 1991.

Tombari, Martin L. *Authentic Testing in the Classroom.* Old Tappan, NJ: Macmillan Library Reference, 1998.

Debunking the Nerd Myth

Glynis McCray
pages 112–115

➲ Additional Resources

Jones, Del, "Stereotype Turns Students Off of High-Paying Career," *USA Today,* 16 January 1998, sec. B, p. 1.

Lewis, Barbara A. *The Kid's Guide to Service Projects: Over 500 Service Ideas for Young People Who Want to Make a Difference.* Minneapolis: Free Spirit Publishing, 1995.

Implementing the National Science Standards

Juliana Texley, Ph.D.
pages 116–123

➲ Footnoted References

National Committee of Excellence in Education. *A Nation at Risk: The Imperative for Educational Reform.* Washington, DC: U.S. Government Printing Office, 1983.

National Council of Teachers of Mathematics. *Curriculum and Evaluation Standards for School Mathematics.* Reston, VA: NCTM, 1989.

National Research Council. *National Science Education Standards,* Washington, DC: Academy Press, 1996.

➲ Additional Resources

American Association for the Advancement of Science, Project 2061. *Benchmarks for Science Literacy.* New York: Oxford University Press, 1993.

American Association for the Advancement of Science, Project 2061. *Science for All Americans* (rev. ed.). New York: Oxford University Press, 1994.

Boone, William J. "Implementation of the Standards: Lessons from a Systemic Initiative." *School Science and Mathematics,* vol. 97, no. 6 (1997).

Bosak, Susan V. *Science is . . . .* Richmond Hill, Ontario, Canada: Scholastic Canada, 1991.

Donlevy, James G., and Tia Rice Donlevy. "Teachers, Technology, and Training: Implementing the National Science Education Standards: Identifying and Removing Barriers to Success for All Students." *International Journal of Instructional Media,* vol. 24, no. 4 (1997).

Goldberg, Howard, and Philip Wagreich. "Focus on Integrating Science and Math." *Science and Children,* vol. 26, no. 5 (1989).

National Academy Press. *National Science Education Standards: Observe, Interact, Change, Learn.* Washington, DC: National Academy Press, 1996.

National Education Goals Panel. *The National Education Goals Report: Executive Summary: Commonly Asked Questions About Standards and Assessment.* Upland, PA: Diane Publishing Company, 1997.

Pathways to the Science Standards: Guidelines for Moving the Vision into Practice, Middle School Edition. Arlington, VA: National Science Teachers Association, 1998.

Stonewater, Jerry K. "The Standards Observation Form: Feedback to Teachers on Classroom Implementation of the Standards." *School Science and Mathematics,* vol. 96, no. 6 (1996).

Applying for Education Grants

Ernest W. Brewer, Ed.D., and Connie Hollingsworth, Ph.D.
pages 124–129

Footnoted References

Brewer, E. W., et al. *Finding Funding: Grant Writing from Start to Finish, Including Project Management and Internet Use* (3d ed.), Thousand Oaks, CA: Corwin Press, 1998.

Additional Resources

Bauer, David G. *The "How-To" Grants Manual: Successful Grantseeking Techniques for Obtaining Public and Private Grants* (3d ed.). Phoenix, AZ: Oryx Press, 1995.

Getting Funded: A Complete Guide to Proposal Writing. (3d ed). Portland, OR: Mary Hall Continuing Education Publications, 1988.

Government Printing Office. *Guide to Department of Education Programs.* Washington, DC: United States Government Printing Office, 1990.

Turning an Educator's Vision into a Classroom Reality

Tamra Ivy
pages 130–137

Footnoted References

Atwell, N. *In the Middle: Reading, Writing, and Learning with Adolescents.* Portsmouth, NH: Boynton/Cook Publishers. 1987.

Gurley-Dilger, L. "Gowin's Vee." *The Science Teacher,* vol. 59, no.3, (March, 1992).

Roth, W. M., and M. Bowen. "The Unfolding Vee." *Science Scope,* vol. 16, no. 5 (1993), 28–32.

Rutherford, F. J., and A. Ahlgren. *Science for All Americans.* New York: Oxford University Press, 1990.

Additional Resources

Bellanca, James, et al. *Blueprints for Thinking in the Cooperative Classroom* (2d ed.). Palantine, IL: Skylight Publishing, 1991.

Flood, James, et al, eds. *Content Area Reading and Learning: Instructional Strategies.* Old Tappan, NJ: Allyn and Bacon, 1989.

Heacox, Diane. *Up from Underachievement: How Teachers, Students, and Parents Can Work Together to Promote Student Success.* Minneapolis: Free Spirit Publishing, 1991.

Jarrett, Denise. *Inquiry Strategies for Science and Mathematics Learning.* Portland, OR: Northwest Regional Educational Laboratory, 1997.

Rathvon, Natalie. *Effective School Interventions: Strategies for Academic Achievement and Social Competence.* New York: Guilford Publications, 1998.

Rhoton, Jack, and Patricia Bowers, eds. *Issues in Science Education.* Arlington, VA: National Science Teachers Association, 1996.

Tomlinson, Carol Ann. *Differentiating Instruction for Mixed-Ability Classrooms.* Alexandria, VA: Association for Supervision and Curriculum Development, 1996.

More Resources!

BOOKS

Armstrong, Thomas. *Multiple Intelligences in the Classroom.* Alexandria, VA: Association for Supervision and Curriculum Development, 1994.

Holt, Larry. *Cooperative Learning in Action.* Columbus, OH: National Middle School Association, 1993.

Mitchell, Ruth, et al. *Learning in Overdrive: Designing Curriculum, Instruction, and Assessment from Standards.* Golden, CO: North American Press, 1995.

Nye, Bill. *Bill Nye's Big Blast of Science.* Reading, MA: Addison Wesley Longman, 1993.

Nye, Bill, and Ian G. Saunders. *Bill Nye the Science Guy's Consider the Following: A Way Cool Set of Science Questions, Answers, and Ideas to Ponder.* New York: Disney Press, 1995.

Schmidt, Mary W. *Teaching Strategies for Inclusive Classrooms: Schools, Students, Strategies, and Success.* Fort Worth, TX: Harcourt Brace College Publishers, 1998.

Schurr, Sandra. *The ABC's of Evaluation: 26 Alternative Ways to Assess Student Progress.* Columbus, OH: National Middle School Association, 1992.

Science and Math Events: Connecting and Competing. Arlington, VA: National Science Teachers Association, 1990.

Science Fairs and Projects, 7–12. Arlington, VA: National Science Teachers Association, 1988.

Scott, John M. *Everyday Science: Real Life Activities.* Portland, ME: J. Weston Walch, 1988.

Scott, Steven A. *Project Connections: Connecting Math and Science Curricula Through Problem Solving.* Pittsburg, KS: Pittsburg State University, 1997.

Stahl, Nancy N., and Robert J. Stahl. *Society and Science: Decision-Making Episodes for Exploring Society, Science, and Technology.* Menlo Park, CA: Innovative Learning Publications, 1995.

Sutman, Francis X., et al. *Learning English Through Science.* Arlington, VA: National Science Teachers Association, 1986.

Teaching Toward Solutions: Step-by-Step Starts. West Nyack, NY: Center for Applied Research in Education, 1998.

Thorson, Annette, ed. *The Guidebook of Federal Resources for K–12 Mathematics and Science,* 1997–98. Columbus, OH: Eisenhower National Clearinghouse, 1997.

U.S. Department of Education. *Excelling in Math and Science: Selected Programs of the U.S. Department of Education.* Washington, DC: U.S. Dept. of Education, Office of Educational Research and Improvement, Educational Resources Information Center, 1995.

U.S. Department of Education. *New Teacher's Guide to the U.S. Department of Education.* Washington, DC: U.S. Department of Education, Office of Educational Research and Improvement, Educational Resources Information Center, 1997.

VanCleave, Janice. *Janice VanCleave's 201 Awesome, Magical, Bizarre, and Incredible Experiments.* New York: John Wiley and Sons, 1994.

VanCleave, Janice. *Janice VanCleave's Science for Every Kid* Series. New York: John Wiley and Sons, 1991.

VanCleave, Janice. *Janice VanCleave's Easy Activities That Make Learning Science Fun* Series. New York: John Wiley and Sons, 1997.

VanCleave, Janice. *Janice VanCleave's Guide to the Best Science Fair Projects.* New York: John Wiley and Sons, 1997.

Vecchione, Glen. *100 First Prize Make-It-Yourself Science Fair Projects.* New York: Sterling Publishing, 1998.

Walpole, Brenda. *175 Science Experiments to Amuse and Amaze Your Friends.* New York: Random House, 1988.

Webb, Willy H. *The Educator's Guide to Solutioning: The Great Things That Happen When You Focus Students on Solutions, Not Problems.* Thousand Oaks, CA: Corwin Press, 1999.

Wiese, Jim. *Roller Coaster Science: 50 Wet, Wacky, Wild, Dizzy Experiments About Things Kids Like Best.* New York: John Wiley and Sons, 1994.

PERIODICALS

Educational Leadership. Published eight times a year by the Association for Supervision and Curriculum Development (ASCD), 1250 North Pitt St., Alexandria, VA 22314; (800) 933-2723.

National Geographic World. Published monthly by the National Geographic Society, P.O. Box 63001, Tampa, FL 33663-3001; (800) 647-5463.

NSTA Reports. Published six times a year by the National Science Teachers Association, 1840 Wilson Blvd., Arlington, VA 22201-3000; (703) 243-7100.

Online Educator. Published monthly on-line by Online Education, 3131 Turtle Creek Blvd., Suite 1250, Dallas, TX 75219; (800) 672-6988.

Quantum. Published bimonthly by the National Science Teachers Association, 1840 Wilson Blvd., Arlington, VA 22201-3000; (703) 243-7100.

Schools in the Middle. Published five times a year by the National Association of Secondary School Principals, 1904 Association Dr., Reston, VA 22091; (800) 253-7746.

Science. Published weekly by the American Association for the Advancement of Science, 1200 New York Ave., N.W., Washington, DC 20005; (800) 351-7542.

Science and Children. Published eight times a year by the National Science Teachers Association, 1840 Wilson Blvd., Arlington, VA 22201-3000; (703) 243-7100.

Science News. Published weekly by Science Service, 1719 N St., N.W., Washington, DC 20036; (202) 785-2255.

Science Scope. Published eight times a year by the National Science Teachers Association, 1840 Wilson Blvd., Arlington, VA 22201-3000; (703) 243-7100.

Science Teacher. Published nine times a year by the National Science Teachers Association, 1840 Wilson Blvd., Arlington, VA 22201-3000; (703) 243-7100.

Scientific American Explorations. Published four times a year by Scientific American Incorporated, P.O. Box, 2053, Harlan, IA 51593; (800) 285-5264.

Teacher Magazine. Published eight times a year by Editorial Projects in Education, 4301 Connecticut Ave., N.W., Suite 432, Washington, DC 20008; (202) 364-4114.

AUDIOVISUALS

Bill Nye, the Science Guy. Video series. 2,080 min. Walt Disney Productions, 500 South Buena Vista St., Burbank, CA 91521; (818) 560-5151.

Just Think: Problem Solving Through Inquiry. Video. 84 min. Office of Educational Television and Public Broadcasting, New York State Education Dept., 1996.

New Assertive Discipline. Video series. 120 min. Canter Educational Products, P.O. Box 2113, Santa Monica CA 90407-2113; (800) 254-9660.

See What Science Is All About. Video series. 14 h. Insights Visual Productions, P.O. Box 230644, Encinitas, CA 92023; (800) 942-0528.

Science Projects and Fairs: Tips for Teachers. Video. 40 min. Insights Visual Productions, P.O. Box 230644, Encinitas, CA 92023; (800) 942-0528.

The Effective Teacher. Video series. 300 min. Harry K. Wong Publishing Co., 943 N. Shoreline Blvd. Mountain View, CA 94043; (650) 965-7896.

Winning Grants II. Video series. 370 min. David G. Bauer and Associates, P.O. Box 6592, Stateline, NV 89449-6592; (800) 228-4630.

ORGANIZATIONS AND ASSOCIATIONS

AIMS Education Foundation
P.O. Box 8120
Fresno, CA 93702
(209) 255-4094
http://www.aimsedu.org

Agency for Instructional Technology (AIT)
Box A
Bloomington, IN 47402-0120
(800) 457-4509
http://www.ait.net

American Association for the Advancement of Science (AAAS)
1200 New York Ave., N.W.
Washington, DC 20005
(800) 351-7542
http://www.aaas.org

American Association for the Advancement of Science: Project 2061
P.O. Box 34446
Washington, DC 20005
(202) 326-6666
http://project2061.aaas.org

American Association of School Administrators (AASA)
1801 North Moore St.
Arlington, VA 22209
(703) 528-0700
http://www.aasa.org

American Association of University Women (AAUW)
1111 16th St., N.W.
Washington, DC 20036
(202) 785-7700
http://www.aauw.org

American Educational Research Association (AERA)
1230 17th Street, N.W.
Washington, DC 20036-3078
(202) 223-9485
http://aera.net

American Federation of Teachers (AFT)
555 New Jersey Ave., N.W.
Washington, DC 20001
(202) 879-4400
http://www.aft.org

Association for the Advancement of Computing in Education (AACE)
P.O. Box 2966
Charlottesville, VA 22902
(804) 973-3987
http://www.aace.org

Association for Supervision and Curriculum Development (ASCD)
1703 North Beauregard St.
Alexandria, VA 22311-1714
(800) 933-2273
http://www.ascd.org

Center for Creative Learning
4152 Independence Court, Suite C-7
Sarasota, FL 34234-2147
(941) 351-8862

Center for Educational Leadership and Technology (CELT)
165 Forest Street
Marlborough, MA 01752
(508) 624-4877
http://www.celt.org

Consortium for School Networking
1555 Connecticut Ave., N.W.
Suite 200
Washington, DC 20036
(202) 466-6296
http://cosn.org

Educational Research Service
2000 Clarendon Blvd.
Arlington, VA 22201
(703) 243-2100
http://www.ers.org

The Eisenhower National Clearinghouse for Mathematics and Science Education (ENC)
Ohio State University
Eisenhower National Clearinghouse
1929 Kenny Road
Columbus, OH 43210-1079
(800) 621-5785
http://www.enc.org

The Exploratorium
3601 Lyon Street
San Francisco, CA 94123
(415) 563-7337
http://www.exploratorium.edu

Federation for Unified Science Education (FUSE)
Capital University
231 Battelle Hall of Science
Columbus, OH 43209
(614) 236-6816

Great Explorations in Math and Science (GEMS)
University of California at Berkeley
Lawrence Hall of Science, #5200
Berkeley, CA 94720-5200
(510) 642-7771
http://www.lhs.berkeley.edu/gems/

Math/Science Network
Mills College
5000 MacArthur Blvd.
Oakland, CA 94613-1301
(510) 430-2222

National Academy of Sciences (NAS)
2101 Constitution Ave., N.W.
Washington, DC 20418
(202) 334-2000
http://www.nas.edu

National Association for Gifted Children (NAGC)
1707 L St., N.W., Suite 550
Washington, DC 20036
(202) 785-4268
http://www.nagc.org

National Association of Secondary School Principals (NASSP)
1904 Association Dr.
Reston, VA 20190
(703) 860-0200
http://www.nassp.org

National Center for Education Statistics TIMSS Project
555 New Jersey Ave., N.W.
Suite 402A
Washington, DC 20208-5574
(202) 219-1828
http://nces.ed.gov/timss/

National Center for Improving Science Education (NCISE)
2000 L St., N.W.
Suite 616
Washington, DC 20036
(202) 467-0652
http://www.wested.org/ncise/

National Center for Science Education (NCSE)
P.O. Box 9477
Berkeley, CA 94709-0477
(800) 290-6006
http://www.natcenscied.org

National Institute for Science Education (NISE)
University of Wisconsin–Madison
1025 West Johnson Street
Madison, WI 53706
(608) 263-9250
http:www.wcer.wisc.edu/NISE/

National Middle School Association (NMSA)
2600 Corporate Exchange Dr., #370
Columbus, OH 43231
(800) 528-6672
http://www.nmsa.org

National Research Center on the Gifted and Talented (NRC/GT)
University of Connecticut
362 Fairfield Road, U-7
Storrs, CT 06269-2007
(860) 486-4676
http://www.gifted.uconn.edu/nrcgt.html

National School Boards Association (NSBA)
1680 Duke Street
Alexandria, VA 22314
(703) 838-6722
http://www.nsba.org

The National Science Foundation (NSF)
4201 Wilson Blvd.
Arlington, VA 22230
(800) 877-8339
http://www.nsf.gov

National Science Resources Center
Smithsonian Institution
955 L'Enfant Plaza, S.W., Suite 8400
Washington, DC 20560-0952
(202) 287-2063
http://www.si.edu/nsrc/

National Science Teachers Association (NSTA)
1840 Wilson Blvd.
Arlington, VA 22201-3000
(703) 243-7100
http://www.nsta.org

National Staff Development Council (NSDC)
P.O. Box 240
Oxford, OH 45056
(800) 727-7288
http://www.nsdc.org

Odyssey of the Mind
P.O. Box 547
Glassboro, NJ 08028-0547
(609) 881-1603
http://www.odyssey.org

Online Education
3131 Turtle Creek Blvd., Suite 1250
Dallas, TX 75219
(800) 672-6988
http://www.learnersonline.com

Project ALERT
Best Foundation
725 S. Figueroa Street, Suite 1615
Los Angeles, CA 90017-5416
(800) 253-7810
http://www.projectalert.best.org

Project CRISS
233 First Ave. East
Kalispell, MT 59901
(406) 758-6440
http://www.projectcriss.org

Project XL
U.S. Patent and Trademark Office
2121 Crystal Drive, Suite 0100
Arlington, VA 22202
(703) 305-8341

School Science and Mathematics Association
Department of Curriculum and Foundations
Bloomsburg University
400 East Second Street
Bloomsburg, PA 17815-1301
(717) 389-4915
http://www.ssma.org

SWEPT
Triangle Coalition for Science and Technology Education
5112 Berwyn Road
College Park, MD 20740-4129
(301) 220-0870
http://www.triangle-coalition.org

U.S. Department of Education
400 Maryland Ave., S.W.
Washington, DC 20202-0498
(800) 872-5326
http://www.ed.gov

Index

G

H

I

J

K

L

M

N

P

Q

R

S

T

U

V

W

We'd Appreciate Your Opinion!

The most valuable feedback about our programs comes from the teachers and students who use them. We invite you to share your thoughts about the *Holt Science and Technology* program by filling out the following questionnaire. This questionnaire can also be found on the *One-Stop Planner CD-ROM,* or you can send us your responses with the touch of a button by accessing the form on the Internet at **go.hrw/surveys/hst.** We would really appreciate your opinion!

Please check the appropriate rating for the *Holt Science and Technology* program as a whole based on the following criteria:

	Excellent	Very good	Good	Fair	Poor
Readability					
Organization					
Ease of use					
Variety of resource materials					
Quality of resource materials					
Your overall rating					

Which features of the Pupil's Edition do you use most? How do you use them?

Which features of the Annotated Teacher's Edition do you use most? How do you use them?

Which features of the Pupil's Edition do you use least and why? Please explain.

Which features of the Annotated Teacher's Edition do you use least and why? Please explain.

Which of the resource materials shown in the chart at right do you use most often? Please place a check in the appropriate boxes.

Resource materials	Use often	Use on occasion	Rarely use	Never use
Directed Reading Worksheets				
Reinforcement & Vocabulary Review Worksheets				
Chapter Tests with Performance-Based Assessment				
Labs You Can Eat				
Whiz-Bang Demonstrations				
Inquiry Labs				
Eco-Labs & Field Activities				
Science Skills Worksheets				
Math Skills for Science				
Critical Thinking & Problem Solving Worksheets				
Assessment Checklists & Rubrics				
Long-Term Projects & Research Ideas				
Holt Anthology of Science Fiction				
Science Fair Guide				

Please explain how you use the resource materials during the school year.

____

____

____

____

Which features of *Holt Science and Technology* did your students enjoy the most? Please explain.

____

____

____

____

Please share any other comments or suggestions you have about the program.

____

____

____

____

Please send your comments to us directly at go.hrw.com/surveys/hst, or mail this form to Holt, Rinehart and Winston, Attn.: HST Feedback, 1120 S. Capital of Texas Highway, Austin, Texas 78746.

Credits

ILLUSTRATIONS

All art, unless otherwise noted, by Holt, Rinehart and Winston.

Page 8 (t), Michael Kirchhoff; 20 (bl), David Merrell; 23 (tr), David Merrell; 40 (bc), Michael Kirchhoff; 62 (t), Michael Kirchhoff; 69 (br), David Merrell.

PHOTOGRAPHY

1, John Langford/HRW Photo; 3, John Langford/HRW Photo; 4, Sam Dudgeon/HRW Photo; 5, John Langford/HRW Photo; 6, John Langford/HRW Photo; 8, Sam Dudgeon/HRW Photo; 9, John Langford/HRW Photo; 10, Sam Dudgeon/HRW Photo; 11, Daniel Schaefer/HRW Photo; 12, Frontera Fotos/Michelle Bridwell/HRW Photo; 13, Sam Dudgeon/HRW Photo; 15, Sam Dudgeon/HRW Photo; 16, Michelle Bridwell/HRW Photo; 17 (br), Daniel Schaefer/HRW Photo; 17 (b), Tim Davis/Photo Researchers, Inc.; 18-19, Tomas Pantin/HRW Photo; 20, EyeWire, Inc.; 21,Peter Van Steen/HRW Photo; 22, Charles Gupton/Stock Market; 23 (tc), Image Copyright © 2001 PhotoDisc, Inc.; 23 (cr), Image Copyright © 2001 PhotoDisc, Inc.; 23 (bl), Image Copyright © 2001 PhotoDisc, Inc.; 24, Victoria Smith/HRW Photo; 28, Sam Dudgeon/HRW Photo; 30, Jack Newkirk/HRW Photo; 34 (tl), Jim Newberry/HRW Photo; 34 (br), Image Copyright © 2001 PhotoDisc, Inc.; 34 (tr), Image Copyright © 2001 PhotoDisc, Inc.; 35, Jim Newberry/HRW Photo; 36, Image Copyright © 2001 PhotoDisc, Inc.; 37 (tr), Sam Lines/HRW Photo; 37 (br), Sam Dudgeon/HRW Photo; 38-39, Sam Dudgeon/HRW Photo; 40, Rader Photography/HRW Photo; 42, Peter Van Steen/HRW Photo; 44 (t), Sam Dudgeon/HRW Photo; 44 (cl), Terry Clark/HRW Photo; 45, Terry Clark/HRW Photo; 46-47, Sam Dudgeon/HRW Photo; 48, Sam Dudgeon/HRW Photo; 50, Bob Wilkey/HRW Photo; 54, Courtesy of Lee Manning; 55, John Henley/Stock Market; 57, Peter Van Steen/HRW Photo; 59, Charles Gupton/Stock Market; 60, Courtesy of Emmett Wright; 61, Corbis Images; 63, Peter Van Steen/HRW Photo; 64 (br), Image Copyright © 2001 PhotoDisc, Inc.; 64 (cl), Ron Pittard/Charting Nature; 64 (bl), Peter Van Steen/HRW Photo; 64 (tl), EyeWire, Inc.; 65 (tl), Peter Van Steen/HRW Photo; 65 (c), Corbis Images; 65 (tr), Peter Van Steen/HRW Photo; 66 (bl), Peter Van Steen/HRW Photo; 66 (t), Roger Neil Photography/HRW Photo; 67, Jose L. Pelaez/Stock Market; 71, PhotoEdit; 72 (tl), Courtesy of Berty Segal Cook; 72 (br), Image Copyright © 2001 PhotoDisc, Inc.; 73, PhotoEdit; 74, Image Copyright © 2001 PhotoDisc, Inc.; 75, EyeWire, Inc.; 78 (bl), Alan Wood/HRW Photo; 78 (tc), Peter Van Steen/HRW Photo; 79, Peter Van Steen/HRW Photo; 81,Peter Van Steen/HRW Photo; 82, Peter Van Steen/HRW Photo; 83, Peter Van Steen/HRW Photo; 84 (tl), Courtesy of John Upham; 84 (br), Courtesy of John Upham; 85 (tl), Courtesy of John Upham; 85 (b), John Langford/HRW Photo; 86 (tl), Sam Dudgeon/HRW Photo; 86 (br), Sam Dudgeon/HRW Photo; 87, Sam Dudgeon/HRW Photo; 88, Ken Pitts/HRW Photo; 90, Sam Dudgeon/HRW Photo; 93, Jose L. Pelaez/Stock Market; 94 (tl), Courtesy of Lloyd Barrow; 94 (bl), Row Lowery/Stock Market; 95, Paul Vismara/The Stock Illustration Source, Inc.; 96, Sam Dudgeon/HRW Photo; 97, Image Copyright © 2001 PhotoDisc, Inc.; 98 (c), Sam Dudgeon/HRW Photo; 98 (tl), Image Copyright © 2001 PhotoDisc, Inc.; 99, Steve Terrill/Stock Market; 100 (tl), Courtesy of Lynn Havens; 100 (tr), Courtesy of Carol Santa; 101, Courtesy of Lynn Havens/Carol Santa; 102, Sam Dudgeon/HRW Photo; 103, Peter Van Steen/HRW Photo; 104, Courtesy of Lynn Havens/Carol Santa; 105, Peter Van Steen/ HRW Photo; 106 (tl), Courtesy of Sandra Schurr; 106 (r), R. Schneider/Image Bank; 107, John Langford/HRW Photo; 110, Peter Van Steen/HRW Photo; 112 (cl), Courtesy of Glynis McCray; 112 (br), Courtesy of Glynis McCray; 113, Courtesy of Glynis McCray; 114, Sam Dudgeon/HRW Photo; 115 (tl), HRW Photo/Sam Dudgeon; 115 (br), Courtesy of Glynis McCray; 116, Marion's Studio/HRW Photo; 117, National Academy Press; 123, Bee Willey/Image Bank; 124 (b), UT Science and Technology Center; 124 (tl), Ferrell Photography/HRW Photo; 124 (tr), Ferrell Photography/HRW Photo; 125, UT Science and Technology Center; 127, Ferrell Photography/HRW Photo; 130, Peter Van Steen/HRW Photo; 133, Ariel Skelley/Stock Market; 134, Peter Van Steen/HRW Photo; 135, Peter Van Steen/HRW Photo; 136, Laurence Dutton/Tony Stone Images; 137, Ron Rovtar/FPG